WELDING
TECHNICAL COMMUNICATION

SUNY series, Studies in Technical Communication

Miles A. Kimball and Charles H. Sides, editors

WELDING TECHNICAL COMMUNICATION

Teaching and Learning Embodied Knowledge

JO MACKIEWICZ

Cover photo of welding instructor taken by the author.

Published by State University of New York Press, Albany

Printed in the United States of America

For information, contact State University of New York Press, Albany, NY
www.sunypress.edu

Library of Congress Cataloging-in-Publication Data

Name: Mackiewicz, Jo, author.
Title: Welding technical communication: teaching and learning embodied knowledge / Jo Mackiewicz, author.
Description: Albany : State University of New York Press, [2022] | Series: SUNY series, Studies in Technical Communication | Includes bibliographical references and index.
Identifiers: ISBN 9781438488516 (hardcover : alk. paper) | ISBN 9781438488530 (ebook) | ISBN 9781438488523 (pbk. : alk. paper)
Further information is available at the Library of Congress.

10 9 8 7 6 5 4 3 2 1

To Ryan Howell, my best welding buddy.

Thank you for the help and the laughs.

Contents

Illustrations

Figures

Tables

Acknowledgments

I am so pleased that *Welding Technical Communication* is part of SUNY's Studies in Scientific and Technical Communication. I'm grateful to Charles Sides and Miles Campbell, the series editors, for seeing promise in a first draft of my manuscript and for encouraging me along the path toward publication. I couldn't have asked for two more supportive editors. I'm also grateful for the guidance of Tim Stookesberry, James Peltz, and Ryan Morris at SUNY Press.

I am lucky that I have a mentor in all things writing related: the author Tony Bukoski. Thank you, Tony, for our conversations about publishing, the English language, and all else.

And, as always, I am grateful to Isabelle Thompson, my dear friend (and frequent collaborator). Along the road to publication, she read drafts of the manuscript and, as always, offered sage feedback.

Of course, I'm particularly indebted to the welding students who participated in the study. And finally, I'd like to thank the welding teachers who allowed me into their labs. Their professionalism and competence continue to inspire me.

Introduction

Welding as Embodied Technical Knowledge

This body is not cooperating.

Apparently, I am incapable of working a vice grip. I am trying to clamp my coupons—two mild-steel plates tacked[1] together on the ends—to the metal table in my booth, but I can't hold the coupons, turn the little tension screw on the vice grip, and hold the vice-grip handles at the same time. To make matters worse, I can barely get my hand around the splayed-out handles. How does anyone do this without four arms? I press the coupons against the table with my belly to give myself a free hand, trying not to brand myself on a still-hot tacked end. For the umpteenth time, the coupons clang onto the cement floor. Trying to pick them up while wearing my welding gloves will be yet another round in this battle. Worse yet, it's summer in central Iowa, and the welding lab does not have air conditioning. I'm getting hotter as minutes slide by, and I become more and more certain that the men in the class are looking at me, thinking I'm pathetic, unable even to get started on practicing my horizontal weld because I can't get the goddamn coupons clamped to the table. This body, my body, does not know what it is supposed to do.

Learning to weld means learning through the body. It means feeling—feeling the most comfortable and stable way to hold the welding gun and feeling the most effective angle and speed. It means learning to breathe as you go. Learning to weld, like learning other skilled trades such as auto and truck mechanics, plumbing, and machining, requires physical as well as mental engagement. It means repeating your efforts and integrating them with abstract concepts until you have learned how to read a situation—a joint, position, metal, temperature, and so on. Haas and Witte (2001) wrote that the development of such embodied knowledge culminates in "the usually skillful and often internalized manipulation of an individual's body and of tools that have become second nature, virtual extensions of the human body" (416). They went on to say that we gain embodied knowledge through "lived experience" until we have a "felt sense" for what works and what does not (417). In brief, embodied knowledge—the kind of knowing that is physical as well as mental—develops through and resides throughout the body as well as the mind.

This book, *Welding Technical Communication: Teaching and Learning Embodied Knowledge* (*WTC*), relates two narratives: one personal and one academic. In regard to the former, I relate the story of how my welding teachers helped me develop embodied knowledge of the technical skilled trade of welding. In this personal narrative, I describe some of my experience as a student at Des Moines Area Community College (DMACC)—a middle-aged woman learning to operate comfortably and effectively amid the tools and talk (including a bit of trash talk) of a welding lab.

I found myself in the welding lab for the first time in January of 2018 after having decided to enroll in night classes at DMACC the December before. I had gone through a spell of feeling sorry for myself, having been diagnosed earlier in the year with labral tears in my hips that made my favorite activities, such as cycling and swimming, uncomfortable if not downright painful. I had waited months for my first surgery, which had been scheduled for December 2017. But two days before the surgery, I got an infection, and the doctor told me the surgery would have to wait. For me, a university professor, waiting meant the surgery wouldn't happen until May—after the spring semester. After two weeks of moping, I decided I simply had to buck up and find something that this body *could* do, something physical. And, for a reason I do not know, what popped into my mind was an old photograph of my grandfather standing among a group of men and women at Globe Shipbuilding in Superior, Wisconsin, my hometown. My grandpa had welded ships during World War II. Learning my grandfather's trade felt right (figure I.1).

In regard to this book's second narrative, I take off my welding helmet and carry out my role as a researcher at Wisconsin Indianhead Technical College in Superior, Wisconsin; Marshalltown Community College in Marshalltown, Iowa; and Lake Superior College in Duluth, Minnesota. For this study, I drew upon research on scaffolded learning theory, technical communication research on embodied knowledge, and

Figure I.1. Joseph Mackiewicz (my grandpa) and others at Globe Shipbuilding in Superior, Wisconsin.

Lave and Wenger's (1991) research on learning within a community of practice. I examined how teachers' verbal communication worked in tandem with their nonverbal communication to build students' embodied knowledge and their moment-by-moment enculturation into a professional community of practice.

Such a study of the technical communication—both verbal and nonverbal—that teachers employ is sorely needed. Although technical communication research has explored the discourses through which people acquire technical expertise (e.g., Fountain 2014), fewer studies have examined how verbal and nonverbal communication combine to support students' acquisition of embodied knowledge. Moreover, about 6.2 million people take classes in two-year colleges in the United States alone (Dougherty, Lahr, and Morest 2017), yet few studies have closely examined the development of embodied knowledge in career and technical education. Even fewer have examined skilled trades such as welding. This gap in the research is a problem because it means that we do not fully understand how teachers' verbal and nonverbal communicative practices—a sort of pedagogical technical communication—scaffold students' learning within the skilled trades.

Scaffolded Teaching with Verbal and Nonverbal Communication

This study of verbal and nonverbal communication in one-to-one pedagogical interactions about welding finds its theoretical basis in research about learning and, in particular, in the concept of scaffolding (Wood, Bruner, and Ross 1976). *Scaffolding* is a frequently employed metaphor used to explain the process of guiding a less-expert other to intended learning outcomes. More specifically, it is the process by which a teacher helps a student accomplish a goal that lies just beyond the student's current capabilities. Scaffolding has (at least) two goals: to help a student succeed in a present-moment task and to help them develop skills and knowledge that will enable them in the future to complete similar tasks with greater ease.

Scaffolded interactions demonstrate six major characteristics. First, for scaffolded learning to proceed, a teacher and student must develop a shared understanding of the task at hand. That shared understanding, in Puntambekar and Hübscher's (2005) terms, is intersubjectivity. Ongoing diagnosis, what Hermkes, Mach, and Minnameier (2018) called dynamic assessment, constitutes the second characteristic of scaffolding. Ongoing diagnosis allows teachers to make their input—their intervention—contingent on (or responsive to) the student's current understanding. Indeed, contingency is the third characteristic of scaffolded teaching.

In a contingent response, a teacher chooses among what have been called *tutoring strategies* (e.g., Cromley and Azevedo 2005; Mackiewicz and Thompson 2018) to intervene in the student's learning. These interventions fall into three overarching categories: instruction strategies, cognitive scaffolding strategies, and motivational scaffolding strategies. With these strategies, teachers give direction, support students in their thinking, and encourage students to continue in their efforts. This study employed a research-based scheme of 12 tutoring strategies to describe welding teachers' scaffolding of students' embodied knowledge and thus their membership in a community of practice. Table I.1 lists the tutoring strategies that I used in this study to describe scaffolded teaching. Of those 12, 11 are verbal. One strategy, demonstrating, is nonverbal.

Related to intersubjectivity, ongoing diagnosis, and contingency is the fourth characteristic of scaffolding: interactivity. Interactivity, which equates to initiating topics and responding to what an interlocutor has said, can be verbal or nonverbal.

Table I.1. The tutoring strategies in this study's data

Category	Strategy	Definition
Instruction	Telling	Teacher directs the student in what to do, using little or no mitigation to lower the face threat of advice.
	Suggesting	Teacher directs the student in what to do, using more mitigation (often negative politeness) to lower the face threat of advice.
	Describing	Teacher relates the characteristics of a thing or action, sometimes with metaphor.
	Explaining	Teacher offers reasons for a given assertion or directive.
	Demonstrating*	Teacher shows how to perform a task.
Cognitive scaffolding	Pumping question	Teacher asks a question that gets a student to respond. Pumping questions vary in the extent to which they constrain a student's response; they can be open-ended or closed.
	Referring to a previous topic	Teacher refers back to the earlier topic or occurrence of an issue.
Motivational scaffolding	Giving sympathy	Teacher acknowledges that the task is difficult for the student.
	Being optimistic	Teacher conveys positivity by asserting a student's future ability to succeed in a task.
	Praising	Teacher points to a student's achievement with positive evaluation. Praise can be formulaic or nonformulaic.
	Showing concern	Teacher builds rapport with a student by demonstrating that they care.
	Using humor	Teacher kids around, tells jokes, or tells amusing stories.

*A nonverbal tutoring strategy.

Scaffolding encompasses two other characteristics. One is called fading. When the student clearly understands the material, the teacher transfers responsibility for learning to the student and leaves the support role. The other characteristic is one that van de Pol, Volman, and Beishuizen (2012) suggested. By checking on a student's learning, they said, teachers determine the extent to which their efforts at scaffolding have worked (203).

This framework of scaffolded learning guided this study of the verbal and nonverbal communication that co-constructed embodied technical knowledge in welding labs. Table I.2 helps to illustrate the differences between verbal and nonverbal communication. Verbal communication refers to language, whether written, signed, or spoken (as in the case

Table I.2. The subcategories of verbal and nonverbal communication.

Mode	**Type**	**Perceived**	**Examples**
Verbal communication	Spoken	Aurally	Tutoring strategies except demonstration
	Written	Visually, tactilely	Text in welding textbooks, on classroom whiteboard; braille script
	Signed	Visually	American Sign Language
Nonverbal Communication	Gesture	Visually	Deictic (pointing) gesticulations
	Paralinguistics	Aurally, visually, olfactorily, tactilely	Syllabic stress for emphasis
	Demonstration	Visually	Showing the procedures for changing a wire spool; showing how to change out a gas canister
	Images	Visually	Photographs and diagrams in welding textbooks; teachers' soapstone sketches on a metal table

of welding interactions). Language differs from other communication in that it exhibits characteristics that other forms of communication do not, including a lexicon and syntax. Like languages such as English, French, and Mandarin, signed languages such as American Sign Language, French Sign Language, and Chinese Sign Language have their own lexicons and syntaxes.

Nonverbal communication comes in many varieties, including images, such as the videos that welding students watched, the blueprints they followed to build test weldments (figure I.2), the photographs that illustrated their texts, and the symbols and icons affixed to equipment (figures I.3, I.4, and I.5). Nonverbal communication also contains the paralinguistic components of language. These are spoken features that communicate but do not change word meaning. For example, tone—pitch that inflects a word—is a paralinguistic feature, but in Mandarin, tone can change word meaning, making it a linguistic component of the language. I discuss paralinguistics in terms of aurally perceived communication such as word emphasis and laughter.

Figure I.2. Blueprint on a student's table. Also on the table: coupons, a tape measure, welpers, and a bottle of Mountain Dew.

Figure I.3. Icons on a welding machine. A welding gun for GTAW rests on top of the machine.

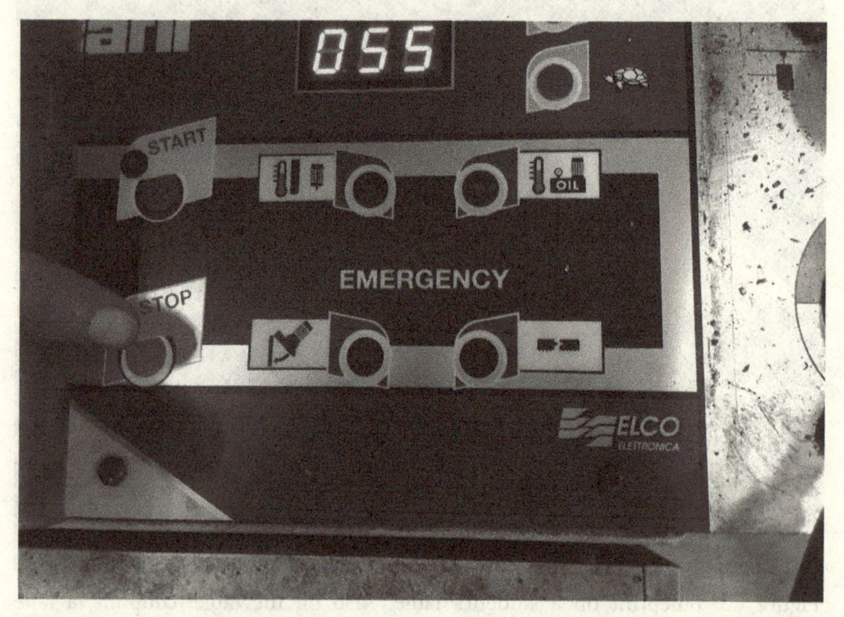

Figure I.4. Icons on a bandsaw. Note the turtle at the top right to indicate a slow speed.

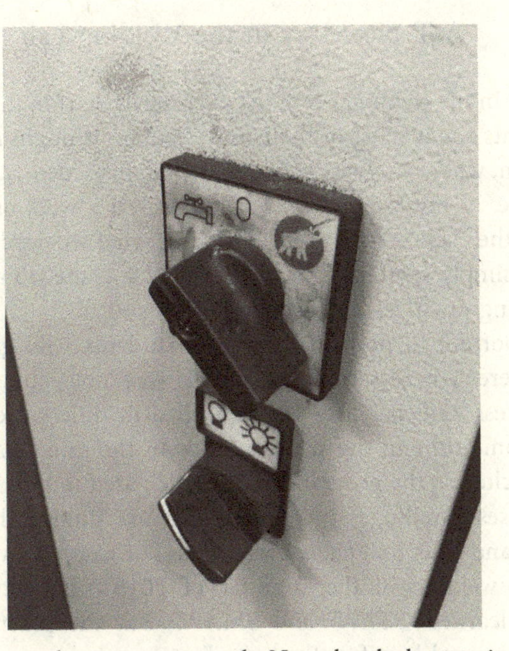

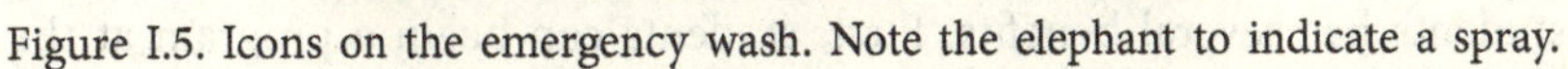
Figure I.5. Icons on the emergency wash. Note the elephant to indicate a spray.

Nonverbal communication is also produced in a variety of ways with the body, including through gesture. Gestures are perceived visually, so in welding interactions, they occur simultaneously with spoken verbal communication. In this study, I analyzed teachers' use of four types of gestures, specifically, gesticulations: iconic, metaphoric, deictic, and beat (McNeill 1992). I analyzed how teachers used these gesticulations in combination with the 12 tutoring strategies listed in table I.1 to scaffold students' embodied knowledge. I discuss and illustrate these gesticulations in detail in chapter 2. But, briefly, iconic gesticulations represent concrete items or actions in the world, whereas metaphoric gesticulations present abstract concepts as if they had concrete form. Deictic gesticulations, in essence, are pointing. And beat gesticulations are rhythmic movements of the hands in time to speech.

As Dix (2016) wrote, "Semiotic systems of language, such as visuals, gestures and actions also scaffold and mediate learning," creating what Sharpe (2006) called "message abundancy." In analyzing these nonverbal elements of communication, my analysis accounts more fully for the co-construction of embodied technical knowledge that occurred during one-to-one welding interactions.

A Typical Day in the Welding Lab

A typical day in the welding lab didn't necessarily start in the lab, the space where students spent most of their time. Rather, it might have started in the classroom, where students gathered to watch a video or, sometimes, to hear a lecture. In my experience, as each semester wore on, class started less often in the classroom. Instead, knowing what we needed to work on, we students simply went straight to our booths in the lab and got started on our welding practice.

It's important to point out that the students who gathered in the classroom weren't necessarily—and often were not—studying the same welding process. Officially, we were enrolled in different classes, but we met at the same time in the same space with the same teacher. Welding programs, including the programs that participated in this study, schedule multiple classes simultaneously mainly because doing otherwise would create small and thus untenable class sizes. For example, when I studied oxyacetylene welding in the summer of 2018, other students in the welding lab learned GTAW[2] and SMAW.[3] But the practice of mixing students has several benefits. First, students can begin a program at the start of any semester in the academic year, and they have more course options available to them during any given semester. More important, at least to me, newer students can learn from more advanced students. In my classes, more advanced students (including students who already worked as welders) frequently helped newer students like me. In my first few semesters, I often sought help from Serge,[4] who worked as a welder with the National Guard. In my later semesters, my friend Sullivan, who had a welding job at John Deere, frequently offered advice and encouragement. And throughout my program, I got ongoing help from my best welding buddy, Ryan, who had worked in welding for years.

Of course, this mélange of students creates a challenge for teachers: Especially in the lab, they must be nimble in their teaching as they move from one process to another, ready to help a student with GTAW and then switch to help a student learning SMAW, for example. However, when I asked him about whether the variety tired him out by the end of the day, Tom pointed out that welders' work experience, including his own, frequently serves as preparation for this: "We were a small job shop, which meant I was switching between processes, techniques, all that stuff multiple times every day. So I was kind of trained to be ready to do that."

Except for the summer semester in which I studied oxyacetylene welding, I'd arrive at school before 6:00 p.m., the start of class. I was usually a little worn out from the workday. During my first few semesters in the program, my classmates were mainly men in their 20s. Before class started, they showed each other oddball YouTube videos on their phones and ate Panda Express and Taco Bell. Sometimes they talked about the drinking they did the weekend before. Sometimes they talked about life in the military. Sometimes they talked about their current welding jobs or jobs that were available. One night I walked in to a job-oriented conversation to hear one of my classmates say, "They can suck my dick for 10 dollars an hour." Such talk went from surprising to routine very quickly. In later semesters, though, the atmosphere in the classroom became less bawdy and more subdued. After a few of the more boisterous students graduated, students tended to sit quietly and look at their phones. In addition, more women started showing up in the classroom, and this demographic change might have made the men in the class less prone to shock talk. In my third semester, another woman student, Samantha, appeared. In semesters thereafter, at least one other woman—and sometimes two or three—were with me.

The labs that I visited differed in their setup, but they shared some features. Each had a large sink for washing up and a first-aid kit (figure I.6). Each had a store of brooms (figure I.7). Welding students are expected to sweep out their booths at the end of class. Each had individual welding booths that lined the walls (figures I.8, I.9, and I.10). Each booth contained a metal table for welding (figures I.11 and I.12). As part of cleaning, each student grinds down the table after class so that it is smooth for the next student. Over the table was some sort of an extraction pipe to remove fumes (figure I.13). Each booth contained a welding machine, made by one of the three main manufacturers: Miller, Lincoln, or ESAB (figures I.14 and I.15). Somewhere in each lab was an "oven" for SMAW electrodes, namely 7018,[5] that needed to be kept from atmospheric moisture (figure 1.16). Usually near that oven was a bank of other electrodes, such as 6010 and 6011 electrodes, as well as filler rods of various metals and sizes (figure I.17). Resting somewhere in the middle of each lab was a large tank of water for quenching and cooling hot metal (figures I.18 and I.19). The labs also contained at least one large metal table that students could use to measure a piece of metal for cutting or to look over a blueprint (figure I.20 and I.21). More commonly, in my

Figure I.6. Sink and first-aid kit at DMACC.

Figure I.7. Brooms for sweeping up after class.

Figure I.8. A row of booths and welding machines at LSC.

Figure I.9. Students in side-by-side booths at LSC.

Figure I.10. Welding booths in the lab at DMACC.

Figure I.11. A welding student in front of his table and stool at LSC.

Figure I.12. A welding table in the lab at DMACC.

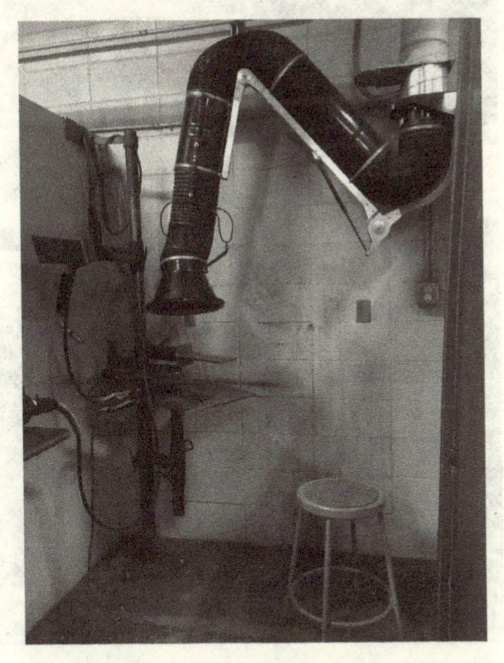

Figure I.13. Ventilation in a booth at NTC.

Figure I.14. Miller welding machines at NTC.

Figure I.15. Miller and Lincoln welding machines at DMACC.

Figure I.16. An oven for 7018 electrodes at DMACC.

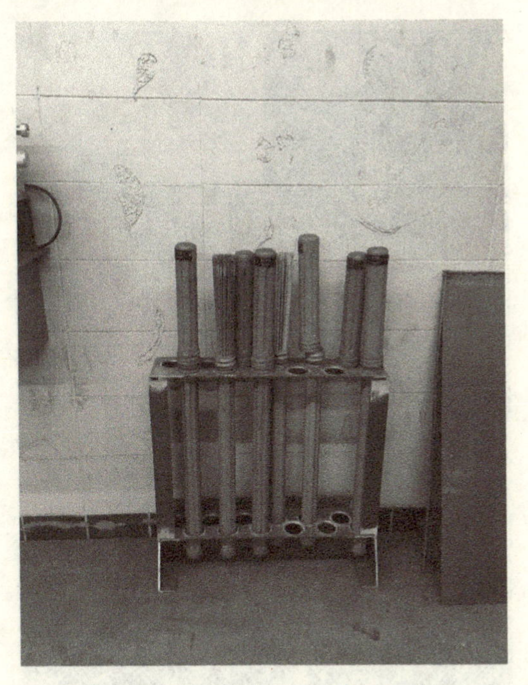

Figure I.17. Canisters of filler rods.

Figure I.18. Quenching bath for hot metal at LSC.

Figure I.19. Quenching bath for hot metal at DMACC.

Figure I.20. Metal tables in the middle of the lab at LSC.

Figure I.21. Metal table in the lab at DMACC.

classes anyway, students gathered around the big metal table, often with newly welded and water-dunked coupons held in welpers,[6] to talk about their work. This ubiquitous metal table is, then, a meeting space.

In managing their time in the lab, teachers tended to move from booth to booth, demonstrating, observing, answering questions, diagnosing problems, helping students set up, and retrieving materials, such as gas tanks, that students need. Tonya said that she had been using this method of touring the lab since her early days as a welding teacher's aide:

TONYA: The first class I ever taught as an instructional aide, the instructor met me at the door. There were 25 students . . . and uh, he said, "Put your hood on, start in this number one, show them what they need to do . . . and we'll meet somewhere in the middle." We met in the middle and he said, "Go back to booth number one. He's doing it all wrong now."

Ted explained his strategy for moving around the lab in more detail:

TED: I try to do every class period, because of course I have 18 students, I try to at least give every student five to six minutes . . . every time. No matter what. Whether I go in there and I watch them. Whether I go in there and I check their weld after they're done. Whether I run a bead for them. Whatever it might be. I try to at least give them that much. Then what I try to do is go back to the ones that need a little bit more time, and I go back to them. Then I come back and say, "Ok. Let's try this. Here's what we need to do."

While teachers tried to rotate systematically through the booths, they frequently got stopped by students with questions. Such a question might lead a teacher to go with the questioner to their booth to diagnose a problem or demonstrate a weld. Or, a teacher might be pulled from their lab circuit to observe some other task, such as a bend test.[7] I'm not sure I ever saw a welding teacher visit each booth in turn without interruption.

The types of one-to-one interactions that teachers had with students in the welding lab fell into four categories. In the first category were interactions that took place when the teacher visited the student's booth on their tour around the lab to check on students' progress. In excerpt I.1, for example, Tom arrived at Sebastian's booth for one of these "making the rounds" checks, greeting Sebastian:

Excerpt I.1[8]

TOM: How we doing Sebastian?

SEBASTIAN: Um, I think I'm doing all right, man. I stopped there [unclear]

Tom assessed the welds that Sebastian had already completed, offered a tentative diagnosis of the problem, then told Sebastian to run another weld while Tom observed:

Excerpt I.1 (continued)

TOM: Yep. And it also looks like you might be running a little bit quick. Let me watch your next bead.

SEBASTIAN: Ok. [Prepares to weld, 19s; welds, 10s]

As Sebastian welded, Tom intervened with instruction, directing Sebastian in how to move his body and the weld gun (figure I.22). Tom's intervention with instruction was directly contingent upon his observation and assessment of Sebastian's technique:

Excerpt I.1 (continued)

TOM: That speed looks good. [2s] Get just a little bit closer to the plate. [6s] Back up a little bit. Stop. [Tom puts his face near Sebastian's weld to inspect it, 18s] I think you might have sped up. Just a touch.

SEBASTIAN: Ok.

After Sebastian finished, Tom looked closely at the weld for 18 seconds, analyzing it and confirming his earlier diagnosis: "I think you might have sped up." Tom then explained how Sebastian's weld bead reflected his slightly too fast travel speed with the gun:

Excerpt I.1 (continued)

TOM: Because you can see how we've got a nice flat bead and then right where it started popping our weld strained. It might have been caused by the popping or it might have

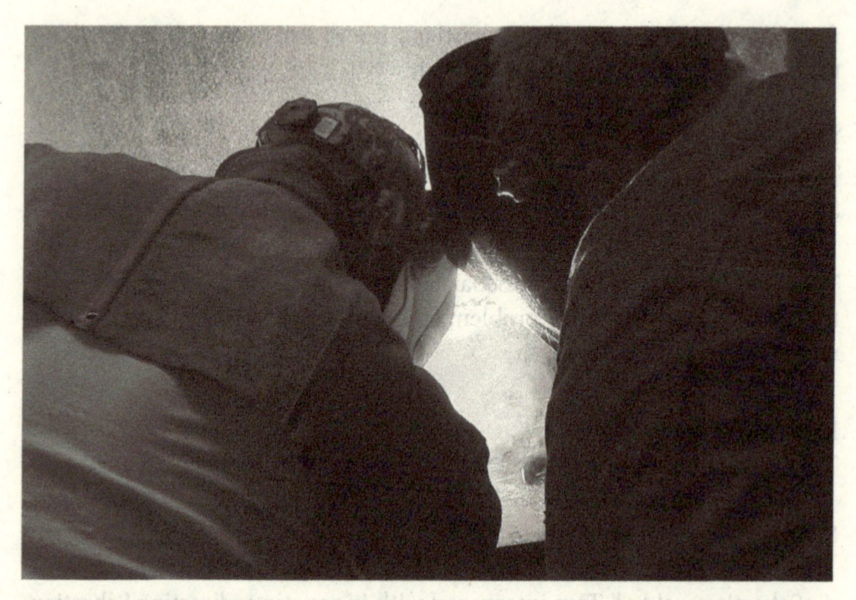

Figure I.22. Tom observed Sebastian to assess Sebastian's technique and, simultaneously, told Sebastian to "Get just a little bit closer to the plate. Back up a little bit. Stop."

Figure I.23. After watching Sebastian, Tom again evaluated Sebastian's work, and then he explained the quality of Sebastian's weld. At the outset of his explanation ("And then right where it started popping, our weld strained") he gestured, pressing his hands together, to convey the concept of strain.

been that we sped up just a little bit and we outran our heat. And that's why it started popping.

SEBASTIAN: Ok.

The interaction in excerpt I.1 was typical of teachers' interactions as they moved from booth to booth in their routine tour of the lab. Tom assessed Sebastian's earlier welds, observed his performance, and assessed the outcome from that performance to offer further ongoing diagnosis. Then, he made his intervention, his explanation of popping and weld strain, contingent upon Sebastian's performance. After this interaction, Tom moved on to the next booth and the process started over again.

A second type of one-to-one interaction that took place in the shared space of the lab, such as near the big metal table, or even near another student's booth, was when a student left their booth to approach the teacher with a question or problem or to display a weldment or some other deliverable for the teacher's evaluation. Excerpt I.2 shows how such an interaction occurred when Stephanie approached Tom just outside of Sebastian's booth with a question. Tom answered her and then built upon that factual knowledge with an explanation for the reason behind his answer. As he explained, he gestured as if sprinkling coarse salt over food (figure I.24):

Excerpt I.2

STEPHANIE: Hey Tom? Tom? Why- what's- what is this?

TOM: That's the silica from inside the wire. The little brown flakes that you see on the normal weld. Because with the spray transfer, you're going to be putting down a lot more weld a lot faster. There's more of it.

After his explanation of the presence of the silica that covered Stephanie's weld, Tom pressed further, asking Stephanie a yes-no question to help Stephanie understand how her actions or, rather, inaction, had likely generated the silica buildup. While he asked the question, he gestured to illustrate the action that she would need to carry out in the future: cleaning the root weld before adding another weld on top of it. He slid his hand back and forth above the length of the weld (figure I.25):

Excerpt I.2 (continued)

TOM: And were you cleaning that off in between?

STEPHANIE: No.

Figure I.24. Standing in front of Sebastian's booth, Tom gestured as he used an explaining strategy to respond to Stephanie's question: "The little brown flakes that you see on the normal weld." He moved his hand as if to sprinkle coarse salt over food.

TOM: No? That's why it's built up so heavy on the top. Because it adds up as you go.

STEPHANIE: So we just like use a wire brush?

TOM: Wire brush. Just a- just a couple of quick passes with a hand brush will clean it off.

STEPHANIE: Ok.

TOM: Yep.

Even before either Tom or Stephanie mentioned a wire brush, Tom's gesture suggested the additional step that Stephanie needed to take and, given the arm movement, even the tool that Stephanie should use to prep the area in order to obtain a good weld. His gesture, then, provided a scaffold that Stephanie could use to answer his question. And, indeed, Stephanie was able to predict the type of tool she should use to clean the root weld. After Tom confirmed her solution, Stephanie went back to her booth to continue her practice.

A third type of interaction took place when the teacher came to the student's booth with a specific pedagogical purpose in mind, usually to demonstrate a technique or, sometimes, equipment setup. These interactions were planned ahead of time. They often occurred when a student would be practicing a certain skill for the first time. Sometimes

Figure I.25. Tom moved his hand back and forth, front to back, simulating the motion of brushing silica off a weld. As he gestured, he asked another yes-no question: "And were you cleaning that off in between?"

the teacher would gather several students to one booth to watch the same mini-class demonstration. In excerpt I.3, Teresa had come to Sophia's booth to demonstrate a horizontal open-root joint in SMAW, a difficult joint for beginners. Before she struck an arc and actually welded the joint, Teresa directed Sophia in the best way to space the coupons apart and tack them. Then, Teresa melded instruction with gestures to clarify the technique that she would demonstrate. Teresa also described an important sound that Sophia should notice and explained its significance:

Excerpt I.3

TERESA: So the whole purpose of this is to be performed between the two plates with a slight drag angle. And what I'm doing is I'm going forward just an eighth of an inch. Coming back and pushing it in. And when I push it in you'll hear a whooshing sound. That whooshing sound is telling me that I'm getting through to the other side.

While describing the proper technique and the sound that would emanate from it, Teresa held the 6010 electrode in her right hand and used it to demonstrate and thus reinforce the dig-and-fill method that Sophia would emulate when it was her turn (figure I.26). As Teresa explained the

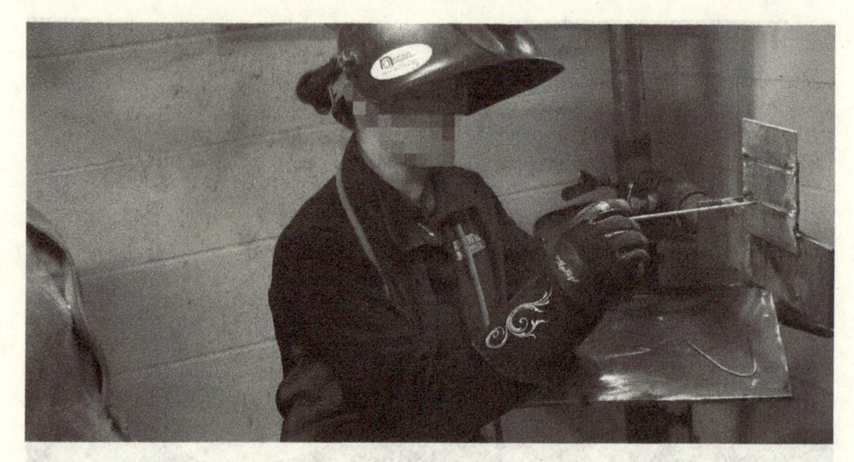

Figure I.26. Teresa demonstrated the dig-and-fill method as she described her movement: "Coming back and pushing it in." Then, she gestured, using the electrode as a pointer.

significance of the "whooshing" sound that would occur if Sophia were to achieve proper penetration in the weld, she pointed to her ear, drawing attention to the importance of this auditory signal (figure I.27). This lesson in welding a horizontal joint went on for several minutes before Teresa struck an arc and performed the weld for Sophia to observe. The entire

Figure I.27. Teresa pointed to her ear as she explained the significance of a sound: "That whooshing sound's telling me that I'm getting through to the other side."

interaction, with its multiple instruction interventions, constituted a sort of introductory lecture on a new joint.

The fourth type of interaction took place when the teacher made a special stop at a student's booth to check on a student because the student had approached the teacher earlier with a question or a problem. The student had to wait until the teacher had time to come to the student's booth. For example, in excerpt I.4, Teresa visited Seth's booth because Seth had approached her earlier about likely gas-induced flaws that he was seeing in his welds. After Seth brought his concern to Teresa as she circulated among the booths, Teresa went outside to adjust the gas running to the student's booth. She then returned to his booth to run a weld and thus ensure that her fix had solved the problem:

Excerpt I.4

TERESA: I turned the argon on, outside.
SETH: Ok.
TERESA: Just the CO_2?
SETH: I turned the four tanks on. They were all hooked up the same.
TERESA: Yeah. They're all- it's all CO_2-
SETH: Yeah.
TERESA: That you turned on. Which is fine because that's what you're using.
SETH: Yeah.
TERESA: But for some reason the system- even though you're using just the CO_2, it has gas in use if you have them both on. So I just turned the argon on and hopefully it's going to run. I mean this one looks pretty good [but it's going to run better
SETH: [Yeah.
TERESA: for you now. I'll just run another one of these and check.

Such visits to a student's booth could relate to specific questions, like Seth's, or they could bring about longer interactions that involved diagnosis and various interventions, such as Tom's instruction in excerpt I.1.

In short, in all four types of one-to-one interactions between welding teachers and students, scaffolded teaching could occur. As the analysis in the chapters to come shows, these interaction types had an effect on the extent to which teachers engaged in ongoing diagnoses and other components of scaffolded teaching.

Conclusion

Welding teachers supported students in their progress toward embodied knowledge and helped them move toward membership in a skilled-trade community of practice. In chapters 3, 4, and 5, I examine this process in more detail, closely analyzing how teachers created intersubjectivity, diagnosed students' knowledge, responded with contingent interventions, facilitated interactivity, faded as students demonstrated greater expertise, and checked on students' progress. To show how they contributed to scaffolded learning, I closely analyze their verbal communication and their nonverbal communication.

Chapter 3 discusses the teaching of three critical components of welding: equipment setup, body position, and technique. It focuses in particular on teachers' instruction, their most directive interventions in students' learning. Chapter 4 covers another critical component of learning how to weld: gaining expert perception—the ability to see, hear, and feel as a welder. This chapter focuses on teachers' cognitive scaffolding strategies, those tutoring strategies that push students to think for themselves. Chapter 5 examines the affective component of teaching newcomers to a field, namely, mitigating students' frustration and fostering a positive environment for learning. This chapter focuses on teachers' use of motivational scaffolding strategies to encourage students to continue in their efforts. Throughout these chapters, I layer my analysis of the role tutoring strategies played in scaffolding students' learning with an analysis of teachers' gestures. Together, these three chapters show how teachers support students through a co-construction of embodied knowledge.

Before these three chapters, though, I first attempt in chapter 1 to articulate the gap in the technical communication research that this study addresses. Then, in chapter 2, I explain the mixed-method approach that I took, combining discourse analysis of teachers' and students' verbal and nonverbal communication with stories from my own experiences as a welding student. In this chapter, too, I discuss women teachers' and students' experiences in welding, including my own.

In the conclusion, I synthesize findings from the chapters, organizing that synthesis around the six characteristics of scaffolded teaching. Throughout, I relate those results to prior research and consider what the findings mean for teaching a technical skilled trade such as welding.

This book arose from my amazement—which persists—that anyone ever learns how to weld. My close analysis of how teachers' and students' verbal and nonverbal communication intertwined to co-construct embodied knowledge is a step toward demystifying that process.

Chapter 1

Situating the Study

Although this study's analysis leans heavily on scaffolded learning theory and research that has tested and developed it, technical communication scholarship, too, greatly informs the analysis of teachers' verbal and nonverbal communication, particularly technical communication research related to the co-construction of embodied knowledge, such as Fountain's (2014) examination of the discourses, multimodal displays, and embodied practices that facilitated students' learning of anatomy in a gross anatomy lab. After all, the verbal and nonverbal communication that welding teachers use to scaffold students' learning is technical communication by any definition of that term. Take, for example, the definition of technical communication on the website of the Society for Technical Communication (STC):

> - Communicating *about technical or specialized topics*, such as computer applications, medical procedures, or environmental regulations.
> - Communicating *by using technology*, such as web pages, help files, or social media sites.
> - Providing *instructions about how to do something*, regardless of how technical the task is or even if technology is used to create or distribute that communication. (STC 2020; emphasis in original)

Each of the components of STC's definition of technical communication relates to the communication that takes place between welding teachers and their students. Certainly, welding teachers communicate about both technical and specialized topics with their students, such as when Tom offered Sebastian an explanation of the results Sebastian had just achieved (excerpt 1.1):

Excerpt 1.1

TOM: Because you can see how we've got, a nice flat bead and then right where it started popping our weld strained. It might have been caused by the popping or it might have been that we sped up just a little bit and we outran our heat. And that's why it started popping.

Tom's explanation helped Sebastian to understand the reasons for the sounds they had heard when Sebastian welded.

While the welding teachers in my study did not communicate via web pages, help files, or social media sites, they did communicate via email and text messages. They also showed videos to their students: videos produced by the Hobart Institute of Welding Technology that demonstrated and explained how to perform various processes, joints, and weld positions, as well as relevant videos on YouTube. They also communicated with technology in a different sense: They used technology, specifically, tools and other implements, to communicate complex ideas or procedures to students. For example, in excerpt I.3 (and figure I.5), Teresa used an electrode to demonstrate the dig-and-fill method that she was teaching to Sophia.

Finally, welding teachers' communication to their students constitutes technical communication in that it includes instruction, in addition to cognitive and motivational scaffolding, intended to help students acquire embodied knowledge. For example, as Tom watched Sebastian weld, Tom told him (i.e., instructed him in) what to do: "Get just a little bit closer to the plate. Back up a little bit. Stop."

Of these three components of STC's definition, the first best encompasses welding teachers' verbal and nonverbal communication. Welding teachers are about the business of helping students—typically newcomers to the trade—understand the technical subject matter of welding. This book examines the ways that they do that—how their verbal and nonverbal technical communication moves students along toward full membership in a community of practice. In that, its purpose fits squarely into what Rude (2009) called the central research question of all technical communication research: *How do texts (print, digital, multimedia; visual, verbal) and related communication practices mediate knowledge, values, and action in a variety of social and professional contexts?* (176; emphasis in original).

Scaffolded teaching is at the heart of this analysis of welding teachers' verbal and nonverbal technical communication. As I mentioned in the introduction, scaffolded teaching manifests six characteristics, apparent in differing degrees in welding teachers' interactions with their students: intersubjectivity, ongoing diagnosis, contingency, interactivity, fading, and checking on students' progress. These characteristics of scaffolded teaching and the learning theory that they represent provide a framework for this study's analysis of welding teachers' verbal and nonverbal communication. This theoretical lens helps to show how teachers tailor their communication so that it moves students forward on the path to competence and in the process moves them forward in their developing membership in what Lave and Wenger (1991) called a community of practice, which they said, "is a set of relations among persons, activity, and world" (98). As they learn and practice, "newcomers" to the community, such as welding students, become more attuned to the community's values, including how and why to perform activities. This process of acclimatization to a community of practice occurs through what Lave and Wenger called legitimate peripheral participation, or LPP. Newcomers interact with "old timers," such as welding teachers, in scaffolded interactions, gradually gaining the embodied knowledge that helps constitute their membership.

Research on communication in workplace settings has tested Lave and Wenger's (1991) theory of LPP. Nielsen (2008), for example, interviewed four apprentice bakers, two journeyman bakers, and one master baker. He also worked for a day at each of the three bakeries where the participants worked. His aim was to study how old timers scaffolded newcomers' learning to create what Collins (1991) called a cognitive apprenticeship: an approach in which "learning is embedded in activities," and these activities "make deliberate use of the social and physical context" (250). Billett (2000, 2001, 2002) was one of the first to closely study the process and efficacy of developing expertise through on-the-job training, and his later research has continued in this vein, often focusing specifically on training in medical professions (e.g., 2015, 2016; Noble and Billett 2017). Across numerous studies, Billett has investigated the role of direct and indirect guidance on workers' performance and development of expertise. Direct guidance occurs via interactions with experienced workers—the old timers—and indirect guidance occurs via encounters with other workplace resources, such as manuals.

Despite the utility of Lave and Wenger's (1991) model, as Fuller and Unwin (2003) pointed out in their analysis of apprentices' developing membership, Lave and Wenger did not account for the ways that formal education, such as education in a welding program, could play a role in students' LPP and thus their progress toward full membership within a community of practice (Fuller and Unwin 2003, 408). Lave and Wenger were more concerned with apprentices' entry into a community of practice as they worked on the job alongside full members, the old timers, of that community.

The present study helps fill this gap in the research by analyzing the verbal and nonverbal communication that comprise welding teachers' scaffolded teaching. In what follows, I examine the prior research that motivated this study. Where possible, I focus on technical communication research, but I draw from a broad range of research relevant to the six characteristics of scaffolded teaching and the learning of expertise.

Intersubjectivity

Intersubjectivity refers to the shared understanding of the task at hand that the welding teacher and student co-construct at the outset of their interaction and maintain throughout. Intersubjectivity includes students' developing ability to perceive phenomena in the environment as an expert. Expert perception develops through study, experience, and practice; it is what welding teachers use to assess students' work, in addition to other phenomena in the welding lab. Through their scaffolded teaching, they help students develop expert perception—the ability to see, hear, feel, touch, and smell as experts. According to Goodwin (1994), teachers "highlight" a specific phenomenon in the environment, making it salient to the learner. Ingold (2001) called such identification of relevant input the "education of attention." Then, teachers "code" that phenomenon, imbuing it with meaning. Discussing anatomy students' development of expertise, Fountain (2014) wrote that through such coding schemes, "a newcomer learns to see as a scientist or expert as he or she comes to embody the practices of that domain" (124). In the context of the welding lab, for example, a welding teacher might point out, or highlight, minute holes in a section of a student's weld—*porosity*. The teacher would then likely explain the meaning of those tiny holes, in this case, the cause of that phenomenon. As teachers scaffold students' perception, a shared understanding—intersubjectivity—grows.

Researchers have drawn attention to the development of expert perception via tactility, hearing, and even smell and taste as well. In addition to describing anatomy students' development of expert vision, Fountain (2014) also discussed the process by which anatomy students learn to make meaning through their bodies, internalizing movements and tactile sensations "as second nature" (100). Other researchers have focused on learning to hear as an expert, such as research on ways music teachers help students hear errors (Thornton 2008) and beats (Miles 1972). More recent research has examined strategies for developing students' tuning perception and performance (e.g., Byo and Schlegel 2016; Silvey, Nápoles, and Springer 2019). Such research suggests, in Silvey, Nápoles, and Springer's (2019) words, "the influential role" of education and practice (393).

Although welding students' ability to smell and, certainly, to taste receive little focus, a discussion of the development of expert perception would not be complete without noting that these senses too, receive attention in other areas of performance, particularly in relation to the taste of wine (e.g., Wang and Spence 2018), beer (e.g., Medoro et al. 2016), and coffee (e.g., Croijmans and Majid 2016). The development of olfactory expertise has been examined in relation to perfumers (Gilbert, Crouch, and Kemp 1998; Livermore and Laing 1996; Zarzo and Stanton 2009) and wine experts (e.g., Brand and Brisson 2012; Mainland et al. 2002). Such research has found, in Royet et al.'s (2013) words, that "in the context of odor experts, it is likely that expertise is acquired with training and experience rather than acquired innately"; in other words, "the notable nose is bred rather than born" (8).

Intersubjectivity also relates to prior research on co-constructed perceptions of objects—both tangible and intangible. Such research has made clear that meanings emerge through shared use and activity. Hindmarsh and Pilnick (2007) revealed the importance of touch, or "intercorporeal knowing," to intersubjectivity by examining a preoperative intubation procedure involving an anesthetist and an operating-department assistant. In the interaction they studied, pressure with a finger on a colleague's hand or the presentation of an instrument provided a way to coordinate a colleague's actions (1408). The interaction participants, they wrote, were "intimately sensitive to delicate and subtle shifts in the embodied conduct of colleagues" (1413).

Focusing on how experts build a shared understanding of tangible objects, such as blueprints, and intangible objects, such as unrealized designs, Sakai et al. (2014), studied plumbers as they planned a pipe configuration, notably recording the meeting from three angles. They noted how plumbers used what they called "situated pointing" to highlight both tangible (via concrete deictic gesticulations) and intangible (via abstract deictic gesticulations) items and thus build a shared understanding of the planning at hand. Besides technical expertise, "shared recognition" (354), they argued, was needed to co-construct the task at hand.

Shared use of inscriptions, such as blueprints, to build a shared understanding was also the focus of Mondada's (2012) analysis of architects' use of plans during a meeting. Mondada wrote that architects ascribed meaning to plans as they read them, folded and unfolded them, leafed through them, and showed them. Thus, Mondada argued that while architects interacted with and through the inscriptions, they also co-constituted them, "voicing absent parties, their validity, and their visibility" (328). In sum, according to Mondada, objects do not exist "out there" (328) but instead emerge through shared use. Similarly, welding teachers and students developed shared understandings of inscriptions, including blueprints, videos, drawings on the classroom whiteboard, and soapstone drawings on the lab's metal table (see figures 1.1 and 1.2), in addition

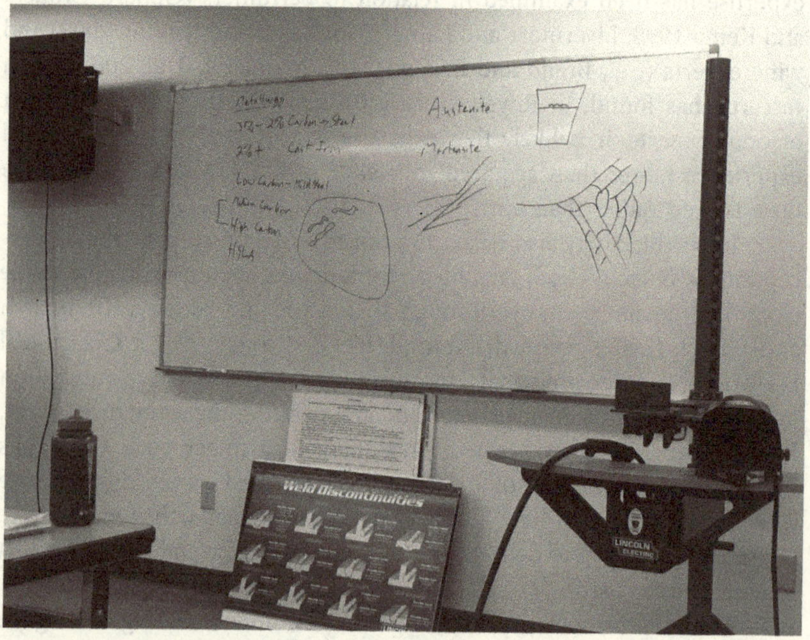

Figure 1.1. Tom's drawings on the classroom whiteboard constituted one type of visual communication that students encountered.

Figure 1.2. Soapstone drawing of a multipass weld on a metal table.

to the intersubjectivity they co-constructed as they examined welds and other student outcomes.

Two studies squarely in the field of technical communication have focused on the role of gesticulations in the development of intersubjectivity.[1] More specifically, they have analyzed how writers use gesticulations during a collaborative writing task. Haas and Witte (2001) analyzed the ways that city employees collaboratively revised a set of engineering standards, including an analysis of the ways that the city employees used gestures as they interacted with the draft document and with each other to construct a shared understanding of the technical standards that would guide city development. They found, for example, that indexical gestures (defined as indicating a real object) and representational gestures (defined as indicating an alternative to current reality) allowed "the city employees to move almost seamlessly from embodied knowledge of antecedent states to embodied knowledge of future states" (444). In addition, drawing from McNeill (1992), they examined literal (i.e., concrete) and nonliteral (i.e., abstract) deictic gestures. In their study, literal pointing referenced a technical drawing or parts of the drawing, and nonliteral gestures referenced objects in the world that the drawing represented (443). They found a difference in the use of pointing by city employees, engineers and nonengineers alike, and pointing as used by consulting engineers. The consulting engineers used less pointing than the city employees did, and when the consulting engineers did use gestures, they tended to use literal pointing gestures, referencing the drawing itself rather than the absent material-world objects with which they were less familiar. Haas and Witte's study revealed the utility of examining gesture in relation to spoken discourse in order to better understand how people develop embodied technical knowledge.

In her case study of a collaborative writing group of technical writing students who worked on a proposal assignment for their upper-division class, Wolfe (2005) developed a coding scheme for the gestures, one that included "adaptor" gestures (Andersen 1999). Adaptor gestures are movements that indicate a speaker's internal state—their anxiety or arousal—and target the speaker themself (as in touching one's hair), objects, or other people. Wolfe included adaptor gestures because "[e]ven movements such as scratching a leg or moving a soda can shape a conversation" and because "such actions draw the gaze and attention of other group members" (305). Basing her coding scheme for gestures on McNeill's (1992), Wolfe found that 60% of gestures were either iconic or deictic, and she found that about 75% of those were generated by just one participant. She also found that all participants used pointing gestures and that pointing alone comprised about 33% of participants' gestures. She found that participants used gestures to write "in the air" in the group's interactional space, to embody the group's document, and to indicate a relationship between the interactional space and the final document. She also found that the participants pointed in order to integrate texts in the interactional space and to maintain a place in the interactional space (309–19). She also analyzed the relationship between gesture and power, as one participant exerted more control on the interaction through his use of gestural space, timing of writing, and noise, such as by clicking a pen (319). Wolfe used Haas and Witte's (2001) elaboration of Witte's (1987) notion of pretext, concluding that "gestures may function as pretexts that help collaborating writers translate emerging, abstract ideas into an embodied representation" (325). Similar to Roth (2002), who studied hands-on activities in tenth-grade physics class, Wolfe found that "by allowing group members to spread their communication across their physical surroundings, gestures reduced the cognitive burden of manipulating internal, verbal representations of the text" (325). In other words, Wolfe, like Roth, found that gestures facilitate the development of intersubjectivity.

Ongoing Diagnosis

As noted above, with ongoing diagnosis, a teacher investigates a student's current level of understanding, and with this diagnosis, shapes an intervention suitable for the student's zone of proximal development, or ZPD. Bounded by the student's current level of mastery and by the student's potential mastery with assistance, the ZPD demarcates a shifting area for growth toward full mastery. In excerpt 1.1, Tom carried out ongoing diagnosis when he observed as Sebastian welded and articulated his assessment of Sebastian's technique ("And it also looks like you might be running a little bit quick").

One way that researchers have examined ongoing diagnosis is by examining students' gestures as a signal of students' level of understanding. Research in the field of L2 learning has investigated how students' gestures, in combination with their verbal communication, help convey their L2 competence and their moment-to-moment thought processes (Gullberg 2006, 2008; Lazaraton 2004; McCafferty 2002, 2004, 2006; Stam 2006). In one such L2 study, Stam (2006) studied English-language learners who spoke Spanish as a first language (L1) to see the extent to which their path gestures (i.e., gestures indicating direction of motion) transferred to English, their second language (L2). In Spanish, path gestures tend to accompany spoken verbs but, in English, path gestures tend to accompany "satellites" of the verb—adverb particles and prepositions (149). Stam found that the Spanish speakers' use of path gestures in English fell somewhere between what would be expected of a native Spanish speaker and a native English speaker, analogous to their English interlanguage. Studies such as Stam's point to the usefulness of analyzing gestures to better understand the development of an L2 and, more generally, to measure learning.

Studying students' learning of a STEM (science, technology, engineering, and mathematics) domain, Scherr (2008) examined how physics students' gestures revealed ongoing changes in the status—from new to familiar—of the ideas that they discussed (5). As Scherr wrote, "Gestures offer one source of evidence of students' engagement in constructive thinking," revealing "scientific ideas that are not yet articulate" (4). Scherr postulated that the students' gestures facilitated their construction of ideas in that they "free[d] up cognitive effort for the task at hand" (7). Scherr's study made clear the potential that analysis of students' gestures holds for assessing students' understanding of complex topics.

Contingency

Based on their diagnoses, teachers determine the intervention that they will provide. McLain (2018), describing a continuum of intervention from restrictive to expansive, put it this way: "The teacher, as a more knowledgeable other, adapts knowledge to be demonstrated to the learner" (988). Van de Pol et al. (2012) described contingency as "tailored adaptation to students' existing understanding" (194), and van de Pol and Elbers (2013) elaborated, "The absolute degree of control does not determine whether scaffolding takes place; it is about the adaptation of the degree of control to a student's understanding that determines scaffolding" (33). As many have discussed before, a teacher's contingent intervention targets the student's ZPD.

Researchers have discussed three types of contingency. Domain contingency equates to determining the topic of focus. At the microlevel, domain contingency relates to the appropriateness of the content or skill the teacher teaches next, given a student's input. At the macrolevel, welding teachers enacted contingency when they organized students' learning at the curriculum and course levels (see also, Gibbons 2002; Smit, van Eerde, and Bakker 2013). In a postinteraction interview, Ted described macrolevel scaffolding of equipment setup when he recalled the instruction that he received and the motivation that experience gave him to teach students how to set up the welding machine themselves, as they will have to do on the job:

TED: One of the things that I had- that really opened my eyes, was my instructor would . . . set the machine up. . . . Well, what did I learn? I learned how to run beads. . . . I didn't know to run the welder. . . . It's just getting them set up. . . . Because no matter what color the welder is,[2] the setup is the same. . . . And I do a basically a crawl, walk, run method when I teach setup. I walk

> you- I basically crawl you right through it. We do it step by step by step exactly the way I want you to set it up. . . . And then I come back the next week, and we do it again. . . . And then now I have you make a list. We make a list in class exactly how we set it up, step by step. You write it down, and now you go in the lab and do it yourself.

Ted scaffolded students learning of machine setup over the course of two or three weeks, first demonstrating how to do it, then further engaging students through their interactivity via writing (list making), then having students enact those lists. Throughout, Ted reduced his control over students' learning, finally fading as students carried out the procedure for themselves. Other examples of macrolevel scaffolding in welding classes are the sequencing of welding positions; students progress from flat, to horizontal, to "out-of-position" welds—vertical down, vertical up, and overhead. Students also move through a sequence of weld joints of different levels of difficulty, progressing from lap joints to T-joints to butt joints. Another example is students' progression through different welding courses, for example, taking GMAW[3] courses before GTAW courses.

Temporal contingency relates to the timing of a teacher's support, or "deciding when to give help" (Rodgers, D'Agostino, Harmey, Kelly, and Brownfield 2016, 346). As Rodgers et al. (2016) pointed out, "Waiting too long before providing help might cause the learner frustration," like my frustration with trying to clamp a coupon to my weld table during my oxyacetylene class. Prior research has examined in depth teachers' positive influence in students' motivation to persist in their efforts. Bolkan and Griffin (2018), for example, examined how teacher behaviors, such as use of humor, caught students' attention and predicted their motivation. On the other hand, "providing help too soon or too frequently may take away important problem-solving opportunities" (Rodgers et al. 2016, 346).

Instructional contingency relates to the extent of the support that teachers give to students based upon diagnosis, from a little to a lot. The extent of support will shift, depending on the student's level of understanding. Much research has focused on instructional contingency—the changing amount of help that teachers or tutors provide as they diagnose students' understanding throughout an interaction. In our study of writing tutors talk, Thompson and I examined how tutors shifted between instruction strategies, such as suggestions, and cognitive scaffolding strategies, such as pumping questions, thereby modulating the extent of the support that they provided (Mackiewicz and Thompson 2018). The present study takes up this focus as well, examining how welding teachers paired, for example, pumping questions and explaining strategies as they scaffolded students' ability to perceive as experts.

As Sharpe (2006) wrote, "other semiotic systems such as visuals, gestures, and actions also act as agents of scaffolding as they help mediate learning" and create a richer message (213). Indeed, a number of studies have examined how teachers have used gesture as a component of their contingent intervention in students' learning. Lazaraton (2004), for example, used McNeill's (1992) typology to examine the gestures of one English-language teacher as she explained vocabulary. She found that the teacher's metaphorical and iconic gestures provided support and redundancy to her verbal messages (106). For example, the teacher used iconic gesticulations to underscore the actions of putting an object somewhere (representing the verb "mislay") and of raising her right hand (representing the verb "swear," as in swear to tell the truth in a courtroom). Building on research such as Lazaraton's, Kim and Cho (2017) examined the frequency with which a writing tutor's scaffolded teaching consisted of spoken language, gesture, or spoken language and gesture together. They found that the tutor most often used spoken language and gestures together. More specifically, they found that the tutor most often used the verbal-nonverbal combination to convey instruction (60% of total gestures) and that the combination most facilitated the learning of students with lower L2 proficiency, especially when helping tutees repair L2 vocabulary and grammar. Kim and Cho's study, too, suggested the important role that gestures can play in teachers' and tutors' interventions in students' learning. In particular, it showed how teachers' gesticulations, especially deictic gestures, promoted "the student's participation and engagement by using writing texts or tools" (114), promoting intersubjectivity in the process.

Research related to teachers' contingent interventions in students' learning has also focused on the metaphors that they use to scaffold students' learning of unfamiliar concepts and skills (e.g., Littlemore and Low 2006a,b; Willox et al. 2010). In her study of metaphor and metonymy in specialized language, Faber (2012) defined metaphor as "the cognitive mechanism whereby one experiential domain is partially mapped or projected onto a different experiential domain, so that the second domain is partially understood in terms of the first" (33). Teachers can use metaphors, as Carter and Pitcher (2010) wrote, to "scaffold learning by using the similarity of vehicle and target systems" (587). The vehicle of a metaphor is the source domain, and the unfamiliar concept is the target domain. For example, as I discuss further in chapter 4, Teresa described the result of pulling the electrode out of the weld joint as an "umbrella of heat." In this case, the source domain was an umbrella and the target domain was the resulting heat that enveloped the joint. Carter and Pitcher pointed out that the "[l]earner's familiarity with the vehicle is crucial, an important consideration when English is not the first language" (588). That is to say, a metaphor is of little teaching value if the learner is unfamiliar with the source domain.

It is worth pointing out that metaphor differs from simile and analogy, as Giles (2008) helpfully exemplified in his thorough examination of metaphors in technical and scientific communication. Giles compared the metaphor "an atom is a solar system" against its simile counterpart, "An atom is like a solar system," and against an analogy, "As the planets orbit the sun, subatomic particles orbit the atom's nucleus" (47). Giles pointed out that metaphor, then, is really "condensed analogy" (47). Approaching the study of metaphor in technical and scientific communication from a rhetorical perspective, Giles noted that Aristotle saw metaphor as a means for "sudden apprehension of similarity, seeing things in a new way" (49) and thus, metaphor helps create knowledge.

To see how social science instructors used metaphors, Low, Littlemore, and Koester (2008) studied three transcribed university lectures to determine, in part, whether metaphors reoccurred or whether they were "one-off" expressions. They found that the most frequently occurring recurrent metaphor was personification, for example, " 'the EU [European Union] is not going to help' " (436). Such personification metaphors occurred in all three lectures they studied. They also found that all three lectures contained multiple examples of nonrecurrent metaphors; these metaphors conveyed their "dictionary meanings," such as "differences that couldn't be whitewashed" and "she comes from the democratic end of the spectrum" (440). In other words, these one-off metaphors were largely conventional, not novel.[4]

Other research on the pedagogical use of metaphors comes from writing center studies. Building off earlier work (Thonus 2010), Thonus and Hewitt (2016) studied writing center tutors' strategic metaphors in asynchronous online consultations. Six writing center consultants received training in strategic metaphor use, and six consultants received no training. The training was "deliberate," they wrote, in that it imparted awareness of metaphor use (59). They found that the tutors who had received training employed metaphors, "possibly deliberately," when responding to students' papers (62). This research suggests that with training, teachers and tutors can use metaphors systematically and coherently.

Interactivity

Scaffolded teaching is necessarily interactive, as the teacher must seek the student's input in order to engage in ongoing diagnosis and contingent response. Indeed, as Lave and Wenger (1991) wrote in their analysis of LPP within a community of practice, "Participation is always based on situated negotiation and renegotiation of meaning in the world" (1520). Melander and Sahlström (2009), for example, showed how three six-year-olds co-constructed an understanding of the size of a blue whale while reading together and how that shared understanding evolved throughout their reading activity. Based on their analysis of the children's conversation, they argued that the changes in the children's shared understanding of a blue whale's size "cannot reasonably be understood as a matter of the expression of changes in individual mental models" (1535). Rather, they argued that we must understand learning, such as the six-year-olds' understanding of blue whales, as relational. In other words, meaning develops through interaction with others, particularly more-expert others. For example, in excerpt 1.2, Stephanie demonstrated interactivity when she inquired about the residue on her weld: "Hey Tom? Tom? Why- what's- what is this?" This view runs counter to, as McCafferty (2002) wrote in his study of gesture in L2 learning, the "Cartesian mind/body split that has so dominated Western thought." He continued, "Cognitive science cultivated a representation of the mind as solipsistic, that is, the individual functioning largely independently of others, and as disembodied as well" (193). Research in scaffolded learning, in contrast, sees interactivity as critical to students' growing understanding in both face-to-face environments (e.g., Murray and Lang 1997; Pratton and Hales 1986) and in online learning environments (e.g., Croxton 2014; Park 2015).

Interactivity is also critical to learning because it leads the student to infer the teacher's meaning and (re)adjust their current understanding. Through this process, called internalization (typically through inner speech), individuals develop control of their own understanding and performance at a certain level (Cazden 2001). Interactivity and the internalization it generates in students demand that they be active, not passive, agents in their own learning. Van de Pol et al. (2012) pointed out that this interactivity works best in facilitating co-constructed knowledge when it manifests openness: "If a student is open about his/her understanding and a teacher is open to the ideas of the student, they can construct new knowledge together" (201).

Much of the research that has examined interactivity in scaffolded teaching has noted its multimodality, including its multimodality via actions and gestures. In a study of air-traffic-control training, Arminen, Koskela, and Palukka (2014) found that both the trainer and the trainee oriented themselves to the time-critical work at hand, creating pedagogical interactions that interwove verbal with embodied action (see also, Hindmarsh, Reynolds, and Dunne 2011; Nevile 2007). Furuyama (2000) focused on gestures in origami teacher-student dyads, looking specifically at the extent to which students' gestures changed in accordance with their teachers' gestures. Furuyama noted what he called "collaborative gestures," or gestures that "interact with the gestures of the communicative partner," such as "the learner's touching and manipulating the parts or the whole" of the teacher's gesture. Indeed, students interacted with "objects" placed in the air by teachers (104). Like Melander and Sahlström's (2009) analysis of children's negotiation of the size of blue whales, Furuyama's study showed that gestures can be collaborative and channels of creating shared understandings.

Fading

Stressing the importance of interactivity to the teacher's successful fading, Many (2002) wrote, "Through dialogue, a teacher is able to orchestrate the shift to self-regulated learning" (379). Via fading, the teacher places more responsibility on the student and withdraws support as the student is able to complete tasks independently. Van de Pol, Volman, and Beishuizen (2010) pointed out that this transfer from teacher to student refers not just to cognitive and metacognitive activities, but also to affect (275). For example, the student becomes responsible for staying motivated and regulating their frustration. Palincsar (1986) connected fading to the aptness of the scaffold metaphor, saying:

> [T]he aim of scaffolded instruction is generalization to less structured contexts requiring less aid. . . . Such generalization is facilitated by the gradual withdrawal of the scaffold as the learner demonstrates increased competence with the task. Given this description, the metaphor of the scaffold becomes clearer because a scaffold is a means of providing support that is both adjustable and temporary. (74–75)

In other words, scaffolds are meant to be temporary and teaching is meant to fade.

In her study of verbal and nonverbal communication in a writing center conference, Thompson (2009) found that the experienced tutor employed a variety of gestures in conjunction with his spoken scaffolding. Simultaneous with a cognitive scaffolding strategy, the tutor used what Thompson called a "fading" gesture, a metaphorical gesture in which the tutor drew his forearms near his body, made eye contact with the student, smiled, and formed "a physical and verbal blank for the student to fill in" (428). Thompson's (2009) study clearly demonstrated the way that gestures can reinforce spoken scaffolding strategies and help enact scaffolded teaching.

In research on workplace training, Filliettaz (2010, 2011, 2013) has analyzed the talk between trainers and apprentices in order to examine the co-construction of embodied knowledge. In research particularly relevant to the current study, Filliettaz (2011) reported two case studies of learning how to weld in the workplace. The first case involved an expert mechanic and an apprentice as they welded a piece of support metal to a cracked chassis. Analysis of their interaction revealed a learning situation that placed the apprentice "in an active role," in which he was progressively given increasing responsibility and was seen as a legitimate partner of a collective work team" (495). This case study, then, revealed in some detail the expert mechanic's fade from the apprentice's learning and the student's progressive LPP in a community of practice.

That said, some research has suggested that fading occurs only rarely (e.g., Mercer 1995). Myhill and Warren (2005) studied fading, or what they called "handover to independence" (59). Specifically, they studied "critical moments," a moment at which the teacher either supports the learner's understanding or hinders it. They studied 54 literacy lesson episodes designed to teach active versus passive voice to two cohorts in primary and middle schools in England. They found that at critical moments, "rather than recognizing or attending to learners' needs" (62), teachers followed a predetermined path toward the learning objective. They concluded that achieving fading, or handover to independence, rarely occurred "because the talk was not set up to facilitate it" (68).

As I discuss further in subsequent chapters, fading occurred in this study's welding interactions, but it did not occur in the tidy fashion described in prior research where the teacher withdraws from helping the student solve a problem as the student progressively demonstrates greater and greater understanding of the task. In a welding lab, teachers do not have sufficient time in a class period to scaffold a student through to mastery on a given task. Fading from an interaction, for this study's welding teachers, often meant leaving the student postintervention to practice on their own.

Checking Students' Learning

In a study of four teachers' scaffolding practices, van de Pol et al. (2012) recommended that checking students' learning be added to the model of scaffolded teaching (203). Thus, I included this last characteristic, manifested in the welding lab when welding teachers, as they made their circuits around the welding lab, checked back with students to assess their progress and, based on that, to provide further (contingent) intervention.

Few studies have examined the impact of checking on students' understanding. However, some education research has studied students' perceptions of teachers' behaviors, including the behavior of checking back with students to gauge their understanding. Garza (2009) studied Latino and White high school students' perceptions of their teachers' caring behaviors and found that the students associated assisting students and checking their comprehension with caring, as they demonstrated "a willingness to ensure student success" (311; see also, Ferreira and Bosworth 2001). In addition, Garza (2009) wrote, this scaffolding practice helps "students who might be timid or embarrassed to ask for help" (312). Indeed, in their case study of training at a vocational school in Geneva, de Saint-Georges and Filliettaz (2008) described how an automotive teacher checked on two students because he heard them striking a metal plate with increasing force—not because the students asked for help. The teacher had previously explained the task, folding a metal plate into a box, and the students had attempted the task on their own, without success (224).

Checking on students' progress as the automotive teacher did makes way for further contingent interventions.

Despite its likely benefits, van de Pol et al. (2012) found that this characteristic of scaffolding occurred infrequently. Writing about their study of three ninth-grade social studies teachers, they said, "A possible explanation for the infrequent occurrence of this step is that the teachers only had one opportunity to practice it. The teachers indicated during the reflection session that they sometimes omitted this step because of time constraints" (202). As I discuss in subsequent chapters, the same problem—time constraints—seemed to impede welding teachers' ability to check back on students' progress.

Conclusion

Welding teachers' verbal and nonverbal communication *is* technical communication; in terms of STC's definition, it clarifies technical information and conveys procedure. It is technical communication, however, within a pedagogical setting. As this literature review shows, prior research has examined learning in similar settings of teaching and learning, such as Billett's (2001, 2002, 2003) studies of apprenticeships in workplaces, Fountain's (2014) study of medical students' learning of anatomy in a gross anatomy lab, and Wolfe's (2005) study of technical communication students' gestures, to name a few. The present study, grounded in a framework of scaffolded learning, continues this tradition, aiming to examine how welding teachers scaffolded their students' learning of the embodied, technical knowledge of welding.

Chapter 2

A Mixed-Method Approach

As I mentioned in the introduction, I used a mixed-method approach—one that employed both discourse analysis of teachers' and students' talk, but also autoethnography. In this chapter, I describe the methods of the study. In the process, I delve into the experiences of women teachers and students, including my own experience as a woman in a welding program.

Autoethnography

Throughout the book, I have related my personal story of gaining embodied knowledge. This personal approach essentially equates to an autoethnography. Autoethnography, according to Ellis (2004), uses self-reflection and introspection to analyze a given context, such as a welding lab. The idea behind autoethnography is that the researcher links the personal to the "cultural, social, and political" (xix; see also, Jones, Adams, and Ellis 2016). A main difference from typical ethnography, then, is that the researcher writes from their own point of view, as opposed to the point of view of others who inhabit the observed culture. In particular, autoethnography uses personal storytelling for sensemaking, a useful tool, I found, for conveying my own development as a welding student as an example of learning and enculturation. An autoethnographic approach, I believe, was appropriate given that reporting my personal experience sheds light on the topic at hand: the development of embodied knowledge through one-to-one interactions.

My status as a woman shaped my personal experience, but not as much as I thought it would. My studies got off to a shaky start, but over the course of several semesters, being a woman in the lab seemed to fade in importance. For one thing, more women—albeit younger than me—enrolled in night classes. As I mentioned in the introduction, my first experience with this newfound companionship came in my third semester of the program. Having taken more courses than me, Samantha was far more advanced as a student than I was. But she was also a far more experienced welder than I was and, indeed, a more experienced welder than most of the men in the class. She already worked as a welder, and she had been welding almost her entire life. Her father was a welder. I found this out when I asked her where she had bought her super-cool leather holder for her welpers. It clipped onto her jeans. She said that her dad had given it to her, so she couldn't point me to the place to get one. (It turns out it's called a plier holster, and you can buy one online.)

After my first two semesters, one, two, or even three other women studied with me. Indeed, as Aghajanian (2018) pointed out, programs in career and technical education have used a variety of incentives to attract women to their programs in an attempt to raise the percentage of welders who are women, which in 2019 was just 5.3% in the United States (US Bureau of Labor Statistics 2020a). We didn't necessarily talk to each other any more than we talked to the men in the lab, but for my part, it felt wonderful to have another woman there with me. One night, for the first hour of class, the only students in the lab were me and the two other women students. The lab felt different to me in that hour—more tranquil. But soon enough Ryan and some other men arrived, and the spell was broken.

And over two years into the program, for my first semester of SMAW, I had my first woman teacher. My SMAW teacher had worked in the industry as a welder for 13 years, and she had been teaching welding at a different DMACC branch for five years. She switched to Ankeny and to night classes so that she could be home with her children during the day. Having a woman as my teacher meant that I would never have to be the only woman in the lab again, and that, to me, was glorious.

Glorious mainly because my first semester at DMACC had been difficult. In that first semester of spring 2018, I enrolled in the safety and blueprint-reading classes that every student has to take before touching any tools or welding machines. My teacher for these first-semester classes singled me out throughout the semester, making fun of my height and my proclivity for taking notes. (See figure 2.1 for an example of my copious note-taking.) His teasing was pretty harmless I suppose, but as the only

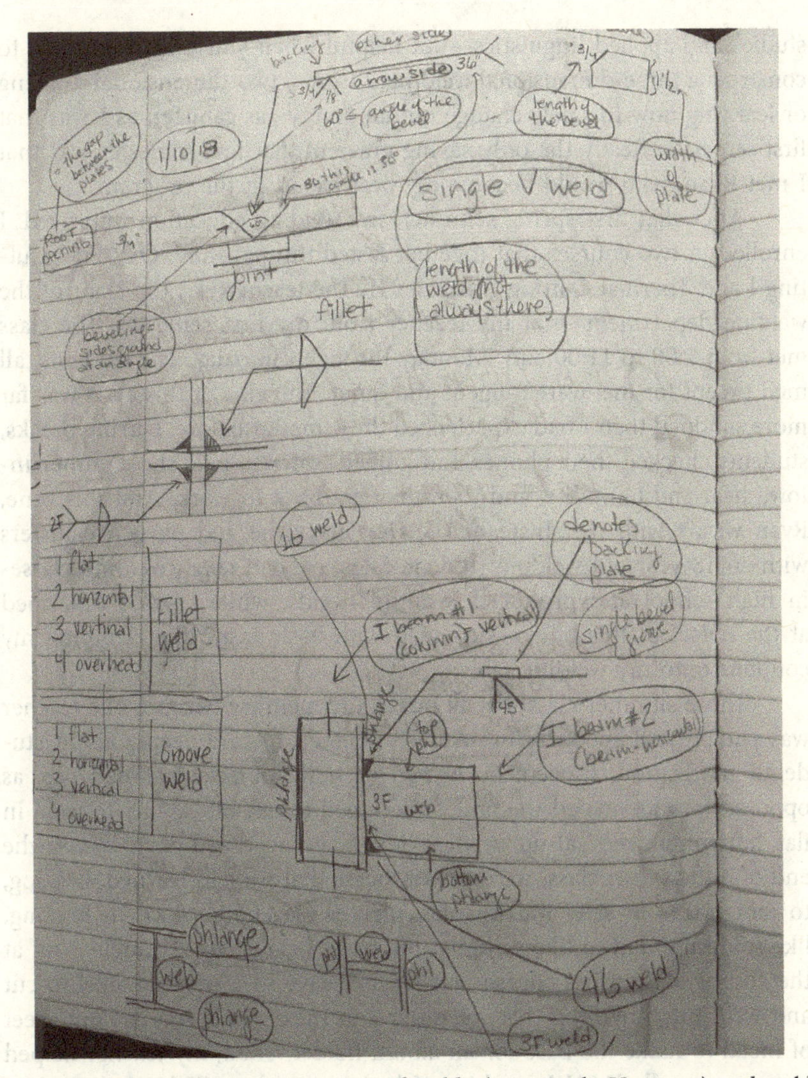

Figure 2.1. My class notes on types of welds (e.g., single V-groove) and weld positions (e.g., flat).

woman in the class and as the oldest person in the class, I already felt like I didn't quite belong. And added on to that, it became quite evident that I didn't have the same background in using tools and being around machines that all the men in the class had. They had military backgrounds and machine-shop experience. I had seven years of grad school in English studies and applied linguistics—not helpful when you're learning how to construct a three-dimensional weldment from a two-dimensional drawing or learning how to safely change out an empty gas canister, as I was that first semester. Really the only saving grace of that first semester was that I met Ryan, who would become my best friend in the program.

After that first spring semester, my welding education improved. I enrolled in two courses that together lasted through July: Oxy-Fuel Cutting I and Thermal Cutting Processes II. The teacher was the chair of the welding department—not my teacher from my first semester. The class met from 7:00 to 11:00 a.m. Monday through Thursday. The students, all men except for me, were diligent and quiet. The vibe of this class was far more subdued than I had experienced the semester before. During breaks, students checked their phones and gulped water to beat the summer-in-Iowa heat and humidity, and then they got back to work. And this time, Ryan wasn't with me. Instead, I shared acetylene and oxygen canisters with an 18-year-old welding student, Silas. He had taken welding classes in high school and produced beautiful welds, while I—as I described at the outset of this book—struggled with basic tasks like clamping my coupons onto my welding table.

Those summer classes stick out in my memory because our teacher was patient and encouraging. And he seemed to care deeply about students' motivation. Indeed, he helped us build objects that had use, as opposed to odd-shaped weldments designed to test our ability to weld in flat, horizontal, vertical-up, vertical-down, and overhead positions. At the end of our oxyfuel class, we used a process that we had learned, brazing, to join pieces of steel into a small planter (see figure 2.2). In brazing, like soldering and unlike welding, the base metal does not melt. And at the end of our cutting course, our teacher gave us a sheet of steel to cut and weld into anything that we wanted to make. I plasma cut my sheet of metal to make the base for an album holder. Then, my teacher helped me use a small GMAW machine (I hadn't yet taken GMAW classes) to weld those strips of metal together and then to weld pieces of pipe to that base. Voilà. A place to keep my Radiohead, Ben Webster, and other albums (see figure 2.3). Throughout that summer, that teacher made me feel like I could succeed, even if I lacked the experience and knowledge that other students already had.

Those first two semesters of my welding education reflect the wide range of experiences that women in welding have encountered. The behavior that my first-semester teacher exhibited led me to think that in

Figure 2.2. The planter that I brazed in the summer of 2018.

Figure 2.3. The album holder that I cut and welded in the summer of 2018.

welding and perhaps all skilled trades, men perceived women as "other," dabblers in the profession rather than potential members. Some of the teachers and students in my study had sensed the same. In an interview, Tom noted that women students had experienced some hardships on the job and in class; and he articulated his approach to mitigating problems in his own classes:

TOM: We've had some females come in and really struggle, and some of that has come from the male classmates. There is sometimes-there is kind of that bias there. I don't know where it comes from necessarily, but sometimes it does present itself. That is something that I always do try and talk to them about as a class too. So it's not just me talking to the women saying, "It's kind of a man's industry right now," and those sorts of things, but also to talk to the male students about that and say, "You know, you're going to run into women out there that are way better than you. Don't discount them."

One student who had struggled in a class was Stephanie. When I interviewed her after recording her interaction with Tom as she learned how to weld around a pipe, she had completed three semesters at LSC. In the interview, she pointed to a very negative experience she'd had when working on a team project with men students:

STEPHANIE: Like two semesters ago, I took a robotics class. And one of the guys in my group literally told me I cannot weld because I'm a girl and would not let me do any of the welding on our project. I actually quit trying in that class and got a D in it. [Laughs sardonically]

When I asked her about the reaction of the other members of the team, she noted that they—all men—did nothing to support her:

STEPHANIE: They didn't really speak up or anything. They kind of just stayed out of it. I don't know.

It's no surprise to find that women are still likely to encounter some obstacles as they pursue their education in a skilled trade. But the story is certainly more complex. For example, when I asked Simon about the presence of women in his classes—whether it made any difference to him—he responded that while women in welding are still rare, he thought it was "great" that they pursued their ambitions:

> SIMON: It is rare, but you know, I guess, it's been considered, probably like a, a guy job to have, but I think it's really cool when girls are actually in there . . . I think it's cool when girls go into what they're interested in. Like Tonya. It's because she loves welding. And you can tell that she's loved it for a long time. And I really like that. Because, um, I don't know, just do what you like to do, you know.

If attitudes like those of Simon, a man in his early 20s, are typical, experiences like the one Stephanie endured in her team robotics project will become rare. In my own classes, besides my bad experience with my teacher's comments during that first semester, women in the lab just didn't seem to be a big deal.

Indeed, I expected Tonya and Teresa—the two women teachers in this study—to share stories of discrimination and other ill-treatment during their time in industry, but that turned out not to be the case. Tonya had been working in welding and welding education for over 27 years when I interviewed her. I thought she would have horror stories about the start of her career, stories of coworkers and bosses who singled her out or treated her as if she didn't belong. Rather, she offered a more nuanced perspective on being a woman in welding.

First of all, she pointed out that the struggle she faced in finding a job as a welder 27 years prior was due, it seemed, to a tough job market, not to her sex:

> TONYA: When I first got into it, I thought that, you know, that it was difficult for me to find work and because I was a woman. . . . But after I started teaching, I learned that it's just as difficult for the men to find a job in the welding industry when they're first starting out. Finding someone to give you that chance. So I kind of changed my attitude- thoughts on that.

When I asked her about her experience in the industry, the treatment that she had received from her men colleagues, she responded that, for the most part, her coworkers had engaged in a sort of benevolent sexism, as opposed to lewd comments or other inappropriate behavior:

TONYA: Yeah, it wasn't real bad. I had- People would tell me that, you know, I was going to hear a lot of foul language and stuff like that, but I didn't experience things like that. I did have a lot of people who were over helpful. . . . It was difficult for the men to see me doing the heavy lifting and such like that. They were always trying to help me. [Laughs]

The story that Teresa told was somewhat similar. While she had worked for a time in auto mechanics for a foul-mouthed, sexist-pig of a boss, Teresa also pointed out that when she worked as a welder, she had few problems with the men with whom she worked and, later, the men that she taught in her classes. They could see that she could do her job, and that was enough for them. Teresa pointed out that, while some men students may have initially questioned her know-how, her ability to "walk the walk and talk the talk" quickly shut them down:

TERESA: Quite quickly, welding is a skill where you can prove that I know what I'm talking about just by physically doing it. . . . Because your students are always going to test you a little bit to see if you actually know what you're talking about, if you can walk the walk and talk the talk.

One of the benefits of welding and other skilled trades for women and other underrepresented groups, then, is that expertise is embodied and, because it is embodied, it is readily demonstrated, verified, and—if justice prevails—accepted. You can't fake it. The knowledge Teresa embodied helped students see her as their teacher, as opposed to their woman teacher.

Teresa's experience jibes with advice given by websites addressing women who are considering a career in welding. For example, the blog site BestWeldingGear warned women that "men might scrutinize your skills and knowledge more than they might for their fellow male coworkers," but then assured them that "if you do great quality work and work hard, then your gender won't matter: you'll just be seen as a great welder" (Celaya 2018). Women who weld have said much the same on Reddit.com. Syraphina (2019), a plumber/pipefitter, wrote, "Overall I haven't had too many issues once I proved myself." Bsd0323 (2019), a welder, wrote, "The dudes I work with are very vulgar but have been everywhere I've worked. Let your welds do the talking." In other words, because expertise in welding is embodied, women can readily demonstrate their qualifications. Even so, I'm sure I'm not the only one who wonders when overcoming men's initial perceptions will no longer be a hurdle for women.

In any case, women increasingly see skilled trades like welding as a way to earn a good living. According to the *Occupational Outlook Handbook* from the US Bureau of Labor Statistics, the median pay for welders, cutters, solderers, and brazers was $42,490 in 2019 (US Bureau of Labor Statistics 2020b). And, as Sophia pointed out in a postinteraction interview, welding made it possible to leave work behind them at the end of the day. Describing her career change to industrial maintenance after nine years in nursing, Sophia said:

SOPHIA: I honestly was just so tired of being so emotionally attached and that it was just draining me so I would take everything home with me. . . . I came here, and like they allow people to shadow classes. And I was just like oh, like that sounds interesting because on my outside time I like to like refurbish furniture and like I've always like wanted to like learn about like electricity and all that stuff. . . . And I came and shadowed it, and they were like building like conveyor belts . . . I was just like, "Wow, this is really cool." And so I was like, you know what? I'm going do it. It pays well.

Sophia dove into a new career despite her lack of shop knowledge. Like me, she came to her first class with little experience with tools. When I asked her whether she had prior experience working with power tools and other shop equipment, she laughed and responded:

SOPHIA: No. [Laughs] I'm always like, "That thing over there," you know. [Laughs] And I mean like I have like my own tools and things but it's just like I don't know their like correct names so I'm just there finding what works, and I know what like does what.

Her (lack of) shop experience contrasted sharply with the backgrounds of the men I interviewed. They tended to be tinkerers, working on farm equipment and cars. Seth, for example, pointed out that he had been fixing machines since he was a kid:

SETH: I've always been like that since I was a kid- trying to fix stuff. . . . When I was younger I guess I used to always have a lot of junk lawn mowers and stuff and I'd always try to get them running or I'd buy junk snow mobiles, four wheelers, junk cars even.

Saul described how his interest in video games led to tinkering with machines and then, eventually, into welding:

SAUL: Definitely it was video games got me into four wheelers and stuff. And then I found out I'm really good at breaking four wheelers. Fixing those throughout the years. And then cars were just naturally the next step. I haven't done a lot on cars, but I screw around. I got into Suburus and kind of brought my dad along with me. Now he's got like four or five.

Historically, women tend to have less shop experience than men, and this deficit can deter women from entering welding and other skilled-trades programs. In reaction, programs in welding and other trades are doing more to ensure that they attract women and provide the resources they will need to enter the industry. For example, Tulsa Welding School offers a Women in Skilled Trades scholarship, as do local chapters of the American Welding Society. In addition, nonprofits such as Women Who Weld, based in Detroit, offer women-only classes and tuition subsidies to attract women. The organization also runs a six-week training program available to unemployed and underemployed women (Aghajanian 2018).

While I started my welding program and this study with the basic idea that women in welding have and will continue to face roadblocks on their way to full membership within welding's community of practice, my research and my own experience have produced a more complicated view. Indeed, this nuance is captured in this comment from Stephanie, who said that other than her bad experience on that robotics-project team, her interactions with men in her classes had been mainly uneventful. But she did point out that she had experienced some condescension:

STEPHANIE: I haven't really had any issues. I mean, I'll have some guys be like, "Oh, that weld is good for a girl." It's like, I don't need to hear that, but ok. But other than that, it's been ok.

Like Stephanie and the other women I talked with, I have had a mostly ok—and in many cases downright positive—experience in welding.

My experience involved coursework toward a diploma in welding. DMACC transferred general-education coursework from my previous undergraduate degree, so I needed to complete only welding courses. Like other welding programs, DMACC's program requires that students spend a certain number of hours in the lab and also that they pass skills tests. DMACC's welding courses are listed in table 2.1.

For the most part, I maintained a student's role at DMACC, as opposed to a researcher's role. However, my welding pals—Ryan, Sullivan, and Serge—knew that I was a professor at Iowa State. They also knew that I was writing a book about welding, but they didn't know the specific topic of the book. Otherwise, none of my classmates seemed particularly

Table 2.1. Welding-program classes at DMACC

Welding course number	**Welding course name**
228	Welding Safety/Health
233	Print Reading/Symbol Interpretation
254	Inspection/Testing Principles
262	Oxy-Fuel Cutting (OFC) I Manual & Mechanized
266	Thermal Cutting Processes II
244	Gas-Metal Arc Welding (GMAW) Short-Circuiting Transfer
245	Gas-Metal Arc Welding (GMAW) Spray Transfer
280	Flux-Cored Arc Welding (FCAW) Self-Shielded
281	Flux-Cored Arc Welding (FCAW) Gas Shielded
274	Shielded-Metal Arc Welding (SMAW) I
275	Shielded-Metal Arc Welding (SMAW) II
251	Gas-Tungsten Arc Welding (GTAW) Carbon Steel
252	Gas-Tungsten Arc Welding (GTAW) Aluminum
253	Gas-Tungsten Arc Welding (GTAW) Stainless Steel

interested in my day job. I think, then, that my experience at DMACC was quite similar to that of other women who make their way through a skilled-trade program.

Discourse Analysis

To elucidate how welding teachers scaffold students' learning of embodied knowledge, this study employed discourse analysis, a linguistic method that concerns itself with closely examining natural language of all modes—spoken, written, or signed. In his germinal work on discourse analysis, Coulthard (1985, 2014) explained that discourse analysis characterizes "how, in the context of negotiation, participants go about the process of interpreting meaning" (viii). Coulthard was arguing that discourse analysis should concern itself with participant meaning-making on a moment-to-moment basis and meaning-making in a broader context, for example, a pedagogical context such as a welding program. While Coulthard's conceptualization of discourse analysis is fairly clear, as Schiffrin, Tannen, and Hamilton (2001) pointed out, pinning down a singular definition of the method is difficult. They noted its "vastness" and "diversity" (5). However, they also noted that various attempts to explain the method include three categories of descriptors. Discourse analysis, they found, applies to the following: "(1) anything beyond the sentence, (2) language use, and (3) a broader range of social practice that includes nonlinguistic and nonspecific instances of language" (1). The present study matches these descriptors of a discourse analysis in that it concerns itself with spoken language beyond the sentence level and accompanying gesticulations with the goal of elucidating scaffolded teaching within a particular community of practice.

It is worth pointing out that discourse analyses, in their focus on broader phenomena, differ somewhat from conversation analyses. Conversation analysis homes in on the details of what people say, including their repetitions, hesitations, overlapping talk, and false starts in order to study conversation "in its own right" (Wooffitt 2005, 18). Conversation analysis focuses on participants' language in order to see how they cooperate to achieve mutual understanding on a moment-to-moment basis. Conversation analyses typically do not, then, look to interviews or other triangulating data to understand the broader context of participants' talk. Because the aim of this study was to understand a teaching context, I gathered and analyzed triangulating interview data.

Teacher Participants

The teachers (two women, two men) who participated in this study were four welding instructors across three welding programs:

- Northwood Technical College (NTC)
- Marshalltown Community College (MCC) in Marshalltown, Iowa
- Lake Superior College (LSC) in Duluth, Minnesota

I interviewed them at multiple points throughout the study. For the informative interviews, I interviewed them before visiting their labs and during downtimes in the lab. For the postinteraction interviews, I called them to ask about the interactions that I had observed. Through these interviews, I learned that their backgrounds in welding and welding instruction varied quite widely.

Tom started college as a mechanical engineering major, but soon changed his mind about his career path:

Tom: There was an introduction to engineering class where you were in a computer lab, you designed a project, so you had a task and then you designed it and then you brought it down to a machine shop and made it. . . . And I remember I went to my advisor after that, and I said, "I don't care if it's in my program. I don't care, you know, where the money comes from to pay for it. I want another class like that where I can actually get in the shop and do something." He just turned to me and said, "You'll have to wait until you're fifth-year engineering." And that was the day I said, "I'm done."

Tom worked for seven years in the industry before starting as an instructor for LSC. He worked first at a small company in Saginaw, Minnesota, as a welder before being laid off. He soon found a job in Superior, Wisconsin, working at a machine shop, using a variety of welding processes. He said about the diversity of work he had done at that shop, "One of the things I tell my students is I remember one day we started at 7:00 a.m. and by noon I had run four different welding processes, three different cutting processes." When I interviewed him for the first time in 2018, he had been teaching for about four years. For one year, he did both, working full time and teaching part time as well. Teaching had shown up "on his radar" after he started tutoring math while in college. The most satisfying part of teaching, he told me, is "getting to watch the students not only develop their knowledge but an actual physical skill as well."

When I interviewed Tonya in 2018, she had been teaching welding for 27 years, the last 10 at MCC. Before that, she taught welding in California. She told me that the job at MCC brought her to Iowa: "They needed someone to build a program here." She got interested in welding when she was tending bar at night. She needed to be home with her children in the evening, and she saw a pamphlet for adult education at a local community college: "And I read about it and I saw- I saw the welding. I didn't even know what welding was." Tonya worked for a brief time in industry after welding school (she was teaching as well), a job that required fabrication, which required skills that she had not acquired in her welding program. She spoke of the foreman of that job as one of the people who had mentored her, eased her way into the community of practice at that small shop: "But then I kind of learned on the job. The foreman taught me. . . . I learned how to use a forklift. . . . My first lesson was, "Get on it. ok, pull that lever. That lifts it. This is that. Ok. You're certified now."

When I interviewed Teresa for the first time in June 2018, she had been working as a welder for over 14 years, and she had been teaching welding for five years at the Superior, Wisconsin, campus of NTC. She started in welding after a bachelor's degree in anthropology and women's studies. After she graduated, she worked in downtown Duluth at a local retail store for about 10 dollars an hour:

TERESA: I didn't feel like I had any useful skills. I read a bunch of books and talked to a bunch of people and wrote a bunch of papers. But I had never done anything. And I really like to travel and I wanted to buy a house and a little farm. And I realized I wanted to make some money to do that. So I figured I'd go for my master's degree. I'd go to the technical school and treat it as my master's degree.

The first degree she earned was in automotive tech. She was unhappy in the job she got after school, but she had learned some welding there, and she decided to go back to school to learn welding. She went to work in the aviation industry for a local company, working three 12-hour shifts per week. That schedule allowed her to accept her welding instructor's offer to work as his teaching assistant. She was working full time as a welder and full time as a teaching assistant when her instructor (now colleague) encouraged her to get her master's degree, which she did through an accelerated summer program that allowed her to earn half of the degree's credits in one summer. She continued to work and teach, but she moved from aviation to shipbuilding. When I interviewed her, she had worked at the local shipbuilding company in Superior, Wisconsin, for about four years.

Ted had worked as a welder for many years when I interviewed him. He spent nearly 20 years in the army, working as a diesel mechanic and a welder. After leaving the army, he worked in the copper mines of northern Michigan for about two years as a welder and mechanic. For a span of 12 years, he worked as a lineman in telecommunications. In 2009, with the Wisconsin National Guard, he served as a welder in Operation Iraqi Freedom. He then worked in waste management as an iron worker for about six years—all before starting as a teacher at LSC, where he had worked when I interviewed him in 2019 for over eight years. During all those work experiences, he continuously supplemented his education, earning degrees in telecommunications and in automotive tech. He also earned a bachelor's degree in career and technical education, and was finishing his master's degree in business when I interviewed him. He told me that the thing he finds most satisfying about teaching is this:

Ted: The end result when you get them students that probably didn't have a whole lot coming into the game. Like I had one student that best job that she had was a nurse's aide. You know, making 10 dollars an hour. Now she's working for the boiler makers. You know, I mean that's huge. When you make somebody's dreams happen, I think that's what it is, you know. That's where teaching comes in. Where you actually make somebody's dreams happen.

These four teachers generously allowed me to follow them around, audio record and video record their interactions with students, and interview them once or more afterward.

Student Participants

The students who participated in the 23 interactions examined in this study (4 women, 15 men) ranged in age from 18 to late 20s. Some enrolled in their programs immediately after high-school graduation. A few of the students who I interviewed—students who had completed internships and were close to graduation—were already working as welders. Others, such as Sophia and Suzanne, returned to school after years in a different career. Four of the students that I interviewed were enrolled in an industrial maintenance program at NTC, as opposed to the welding program. Industrial maintenance programs tend to require at least one course in welding. Like maintenance students at DMACC, these students move through all the processes and positions but do so at a much accelerated pace, tackling a degree's worth of welding competencies in one semester.

Data Collection and Transcription

Getting permission to conduct the study. For this study, I submitted two applications to conduct research involving human beings. The first application was limited to conducting interview research, which I began in April 2018. The second application requested approval for a study that would video record and audio record welding interactions and audio record interviews and subsequently analyze those recordings and transcripts. I received institutional review board (IRB) approval from Iowa State for both studies. In addition, I received permission to conduct research from the administration at MCC[1] and from the IRBs at LSC and NTC. I began setting up visits to the three schools soon after receiving permission in October 2018. Because I had already met two teachers through the prior interview study, I was able to set up visits for the following month.

Finding participants for the study. The teachers of the welding classes helped me obtain student participants for the study. They emailed the study's consent form (appendix C[2]), attached to a solicitation message (appendix D), to students before I showed up to class. When I arrived, the teachers introduced me and gave me a few minutes to talk to the class about the study. Some students were adamant that they did not want to participate, but others told me that I was welcome to record them at any point during the class period. And those students also agreed to a postinteraction interview, either face-to-face or by phone.

I had met Tom and Teresa prior to the start of the video recording. I met Teresa by chance while sitting in a coffee shop. I asked her whether I could contact her later to talk, and we met at the coffee shop again for an hour-long informative interview. The next semester I began video and audio recording in her classes. I met Tom by emailing him and introducing myself. As with Teresa, I interviewed Tom for an hour in the semester before I attended his class to video and audio record interactions. I also visited his class, listening to a test-review lecture and then observing the students in the lab. I met Tonya through an introductory email. She agreed to participate in the study and to send out my solicitation email to her students before I arrived at her class to video record. Similarly, I met Ted through an introductory email, but in this case, I was able to point to my acquaintance with Tom, also a teacher at LSC. Ted agreed to participate and to send out my solicitation email to his students. Besides visiting his lab to record welding interactions, I also visited his classroom, listening to a lecture on proper use of a grinder.[3]

Video and audio recording interactions. I video recorded 23 interactions across the three schools. In cases where I was able to plan a few minutes ahead, I also audio recorded the interactions. I was able to audio record as well when the teacher would tell me who they were planning to visit next as they traveled around the welding lab. In these cases, I visited the student's booth in advance and set up my digital audio recorder. The audio recording provided a backup to the video and, in some cases, provided a better-quality audio recording of the interaction. In other cases, though, I found interactions to video record by trailing the teacher around the lab as they checked in on students or visited students who had asked for help. In these cases, I asked the student for permission to video record and, if they said yes, and went through the consent form with them. In cases where I could not get the audio recorder set up before the interaction, sometimes I held the video recorder in one hand and the audio recorder in the other. Redundancy in recording was critical because (1) many processes, such as carbon-arc gauging, are extremely loud; (2) participants wear welding helmets, or hoods, which muffle their speech; (3) the recording equipment has to be kept at a relatively safe distance from the flame or arc and thus away from the participants.

Transcribing. Two research assistants transcribed some of the audio recordings of interviews. I also used REV.com, an online transcription service, for the interview data. In contrast, transcribing the one-to-one interactions was much more difficult because of the difficult setting and thus the oftentimes challenging sound quality. Plus, some knowledge of welding is helpful when transcribing welding interactions given that specialized terms abound. Thus, I transcribed the welding interactions myself. I used orthographic transcription. In addition to spoken words, I transcribed the following extralinguistic features:

- Unintelligible talk, with "unclear" in brackets, as in [unclear].
- Laughter, with "laughs" in brackets, as in [Laughs].
- Actions, with relevant action in brackets, as in the following example:

TOM: [Sebastian welds, 10s] That speed looks good. [2s] Get just a little bit closer to the plate. [6s] Back up a little bit. Stop. [Tom puts his face near Sebastian's weld to inspect it, 18s] I think you might have sped up. Just a touch.

- Pauses longer than one second, with the number of seconds in brackets, as in [2s].
- Pauses one second or less, with a comma.
- Rising intonation for an inquiry, with a question mark, as in the following example:

TERESA: You see these little rivers?

- Cut-off speech, with a hyphen, as in the following example:

TONYA: Yeah, that's a- It would be a good idea for you to go over those chapters.

- Reference to a word as a word, with double quotation marks, as in the following example:

TERESA: It's called a "keyhole."

- Occurrences of overlapping talk, denoted with brackets as in the following exchange:

TERESA: When you go to that little W motion it'll help flatten it out here [So
SUSAN: [Yeah

TERESA: you went a little slow right here [but most of it looks good
SUSAN: [Ok

I also ensured that utterances such as "uhhuh," "ok," and "mmhm" were spelled consistently. For example, "uhhuh" became the consistent spelling for "uh-huh" and "uh huh." Also, "ok" became the consistent spelling, as opposed to "okay" or "OK." And I used "mmhm" instead of "mmhmm" or "mhm."

THE DATA: WELDING INTERACTIONS AND INTERVIEWS

Welding interactions. As mentioned previously, I video recorded 23 welding interactions at the three schools. The interactions averaged around five minutes. However, this average is somewhat deceiving; in several cases, silence punctuated the interactions as the teacher demonstrated a process, speaking only occasionally. In addition, interactions sometimes occurred in two parts—one part might contain a demonstration or analysis of the student's work and the other part might contain a check to see whether the advice helped.

Interviews. After I video recorded each welding interaction, I contacted the student participant to set up a postinteraction interview. In some cases, I could not get a response from the student participant. In total, I audio recorded eight 10- to 15-minute postinteraction interviews. In these interviews, I asked the students about their educational background, their experience with welding, their reasons for entering a welding program, their preferences for welding instruction, (for men) their thoughts about women joining welding programs, (for women) their experiences as women in welding programs.

I also conducted both informative and postinteraction interviews with the instructors. In the informative interviews, I asked teachers about their experiences in welding and welding instruction, as well as some of their teaching methods. In postinteraction interviews, I asked them about the interactions that they had just had with students. For example, I asked about specific methods that they had used in their one-to-one interactions. However, I also asked them broader questions, aiming to better understand their experience as welders and welding instructors.

The postinteraction interviews were semistructured; I brought a set of questions to them (see appendices A and B), but I deviated from those questions and followed up on participants' answers as necessary.

Most important, I asked participants for their perceptions of their specific one-to-one interactions. That said, because it was not possible to have the welding interactions transcribed and analyzed in time for the interviews, I could not carry out stimulated-recall interviews, asking participants about specific utterances, such as specific tutoring strategies, or specific gestures. A stimulated-recall method like this would have been informative; the problem with stimulated recall as a method, however, is the difficulty of implementing it. In a stimulated-recall interview, the researcher has to play back the recording, note particular items of interest, and schedule an interview with the student or teacher as soon as possible after the initial interaction—preferably within 24 hours. But students have other classes and jobs, and teachers have other classes to teach.

Discourse Analysis of Welding Interactions

As described in the introduction, I analyzed the welding interactions for characteristics of scaffolded teaching, including ongoing diagnosis, contingency, and interactivity. More specifically, I examined how teachers intervened in students' learning through a range of tutoring strategies. I used a framework first developed by Cromley and Azevedo (2005) through analysis of adult literacy tutoring. Mackiewicz and Thompson (2018) modified this scheme for use in analyzing writing tutor-student writer conferences, delineating 16 reliable strategies through multiple rounds of revision and interrater testing. I modified the Mackiewicz and Thompson framework in four ways:

- Adding a strategy apparent in the welding-lab context that the original framework failed to capture. Specifically, I added the describing strategy.
- Removing strategies that did not occur in the welding-lab context.
- Separating strategies that were merged in the original framework but more usefully considered separately in the welding-lab context (e.g., being optimistic and using humor).
- Moving a strategy from one category to another. Specifically, I moved the strategy of demonstrating from the category of cognitive scaffolding to the category of instruction. Mack-

> iewicz and Thompson (2018) classified demonstrating as a cognitive scaffolding strategy. I reassessed this strategy for the context of welding interactions, where the teacher demonstrates an action with the intent that the student will eventually (usually within a few minutes) imitate that action (or avoid it, in cases where the teacher showed the student what *not* to do).

In the end, I analyzed welding teachers' scaffolded teaching with a scheme of 12 tutoring strategies, 11 verbal and one visual (i.e., demonstrating).

Analysis of Gestures

The visual communication mode of gesture is inseparable from spoken communication. As McNeill (2016) wrote, "*the core is gesture and speech together.* They are bound more tightly than saying the gesture is an 'add-on' or 'ornament' implies. They are united as a matter of thought itself" (3; emphasis in original). McNeill's body of work on gesture (e.g., 1992, 2005, 2016) developed the idea that our gestures constitute an essential component of our communication. Researchers such as McNeill (1992, 2005), but also Kendon (2000, 2004) and Kita (2000), have argued that gestures enable spatio-motoric thinking, the type of thinking that coordinates information with action and occurs as people use their bodies to engage with their physical environments. In contrast, analytic thinking hierarchically structures information into "conceptual templates" and underlies language (Kita 2000, 164). In this study, analyzing gestures along with verbal communication helped to reveal the range the communicative resources through which teachers and students co-constructed embodied technical knowledge.

My analysis of gestures followed McNeill's (1992, 2005) frequently used delineation of gesture types, which McNeill called "Kendon's Continuum," after the gesture scholar Adam Kendon. This framework includes the categories of gesticulation, language-like gestures, pantomimes, emblems, and sign language:

- Gesticulations are idiosyncratic and spontaneous and almost always combine with spoken communication. Gesticulations comprise four types: iconic, metaphoric, deictic, and beats. I discuss and exemplify these types in more depth below.

- "Language-like" gestures take the place of a word or phrase (i.e., some grammatical slot) in verbal communication. For example, a reporter could say to the cameraperson, "I can't ________," pointing to their ear to indicate "hear" rather than speaking the word.

- In general, pantomimes occur in the absence of spoken communication, including spoken verbal communication. Indeed, they are often used in contexts in which speech is not an option (van Nispen et al. 2017). With pantomime, people act out a thought.

- Unlike pantomimes and language-like gestures, emblems, such as the "ok" sign, have "standards of well-formedness" (McNeill 1992, 37). They must be performed in a certain way to convey their meaning. Emblems include praise, such as the "thumbs-up" sign (at least, in the United States) seen in figure 6.13 and insults, such as "flipping the bird" (again, in the United States, at least).

- Sign language is true language, with all the characteristics of language, such as arbitrariness and syntax (McNeill 1992, 38–39).

My study focused on gesticulations, which occurred more frequently by far than the other types of gestures in welding interactions as they combined with teachers' spoken communication in their scaffolded teaching.

Before focusing more specifically on gesticulations, I want to distinguish sign language from the other four categories of gestures, including gesticulations. Unlike the other types of gestures, the gestures of sign languages generate true language. Like other languages, they manifest these characteristics:

- Segmentation, meaning that segments compose meanings.
- Compositionality, meaning that combining segments generates meanings.
- A lexicon, meaning that individual segment forms recur in different contexts.
- A syntax, meaning that segment combinations adhere to standard patterns.

- Distinctiveness, meaning that details added to segments distinguish them from other segments.
- Arbitrariness, meaning that signs are not iconic.
- A community of users, meaning that some community understands the signs and their combinations.
- Standards of well-formedness, meaning that certain standards apply for a particular sign to be considered that sign (McNeill 1992, 38–39).

In this study, neither the welding teachers nor the student participants employed a sign language, such as American Sign Language. And, when I asked the welding teachers whether they had any knowledge of or experience using a sign language, they all confirmed that they did not.

On the Kendon's Continuum from gesticulation to sign language, "(1) the obligatory presence of speech declines, (2) the presence of language properties increases, and (3) idiosyncratic gestures are replaced by socially regulated signs" (McNeill 1992, 37). The least language-like of all gesture types, gesticulations differ from sign language in that they are

- global and synthetic. McNeill wrote that "the meanings of the parts of a gesture are determined by the whole" (41).
- noncombinatoric. "Gestures," he wrote, "don't combine to form larger, hierarchically structured gestures" (41).
- lacking standards of form. "Different speakers display the same meanings in idiosyncratic ways. . . . [I]ndividuals create their own gesture symbols for the same event" (41).
- context sensitive. "Each gesture is created at the moment of speaking and highlights what is relevant, and the same entity can be referred to by gestures that have changed their form" (41).

This fourth bullet point—the context sensitivity of gesticulations—made analyzing teachers' gesticulations critical. Teachers' gesticulations helped generate their contingent interventions, such as instruction, in students' learning. And, as I pointed out in chapter 1, although technical communication research has examined in depth the role of visual communication in the form of data displays (e.g., Kostelnick 2016; Li 2020; Rawlins and Wilson 2014; Wilson, Rawlins, and Crane 2018), comics (e.g., Bahl, Figueiredo, and Shivener 2020; Yu 2016), and technical illustrations (e.g., Alexander, Schubert, and Meng 2016; Ross 2017; Zhang 2016), it has with just a few exceptions overlooked the role of gestures in communicating—specifically teaching—technical knowledge.

As noted above, gesticulations comprise four types: iconic, metaphoric, deictic, and beats. Iconic gesticulations represent concrete items or actions in the world. Figures 2.4 and 2.5 show Ted using an iconic gesticulation to help describe, in Ted's terms, "a couple hangers that went off the track." Ted moved his hand in a slicing U motion, mimicking Sawyer's welds that had gone askew.

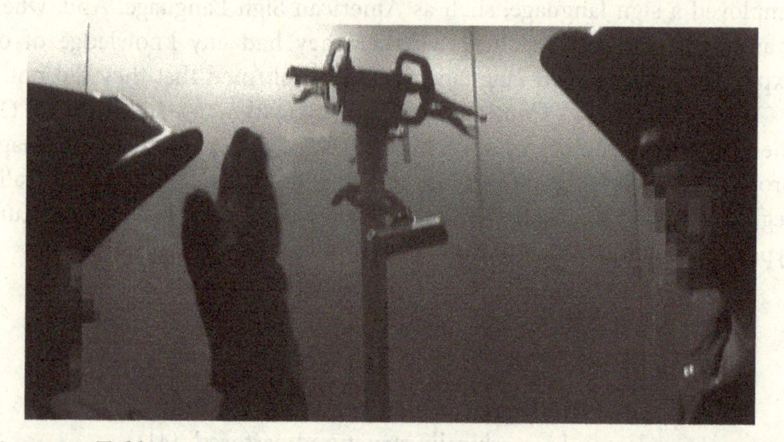

Figure 2.4. Ted began to move his hand to make a U, using an iconic gesticulation as he noted "a couple hangers that went off the track."

Figure 2.5. Ted moved his hand in to make the second half of the U, finishing the iconic gesticulation as he noted "a couple hangers that went off the track."

In contrast, metaphoric gesticulations present abstract concepts as if they had concrete form. The concept, or the target domain, is conveyed visually as a physical object (McNeill 1992, 14–15). This physical object is the source domain (see also, Kövecses 2010; Lakoff and Johnson 1980). Figure 2.6 shows Tonya using a metaphoric gesticulation to refer to the abstract concept of temperature, specifically, the shift from a lower temperature to a higher temperature as Shawn gained experience in the vertical-up position.

Deictic gesticulations are what McNeill (2005) called "an extended 'index' finger," but he noted that people could use other body parts or any held object (such as a tool) (39). Indeed, Kita (2003) argued that pointing "is a uniquely human behavior" (2) that "is one of the first versatile communicative devices that an infant acquires (2). Deictic gesticulations carried out two types of pointing. Concrete pointing, or concrete deixis, located an object or an action in space in relation to a reference point (McNeill 2005, 39–40). Figure 2.7 shows Tonya engaging in a concrete deictic gesticulation when she pointed to the edge of Steve's lap joint.

In addition, McNeill (2005) discussed abstract deictic gestures (see also, Gullberg 1998), which point to locations in space in order to refer to abstract ideas. Figure 2.8 shows Tom performing an abstract deictic gesticulation as he showed Sullivan how to use a plasma cutter to cut and bevel a section of pipe. Tom pointed away from the section of pipe that

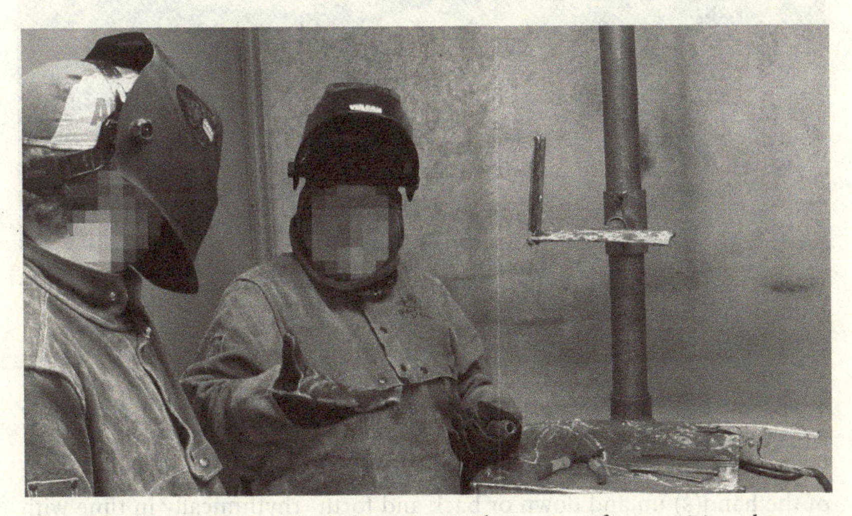

Figure 2.6. Tonya used a metaphoric gesticulation to refer to raising the temperature, the amperage, and the voltage, as Shawn progressed. "Then you want to start to take that temperature up."

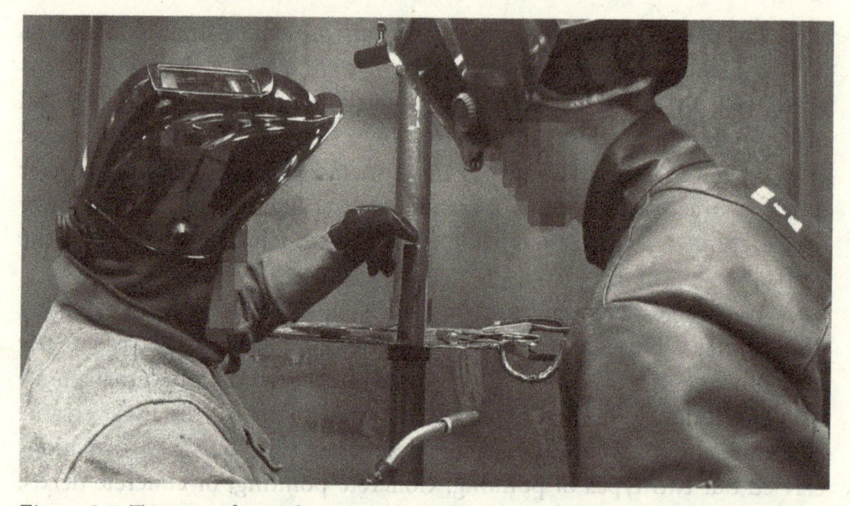

Figure 2.7. Tonya performed a concrete deictic gesticulation, in this case, pointing to the edge of a lap joint.

Figure 2.8. Tom used an abstract deictic gesticulation as he pointed to a possible future section of pipe for practice.

he and Sullivan had just cut—the present object—and toward the future, as-yet-nonexistent pipe.

The fourth type of gesticulation, beat gesticulations, are "mere flicks of the hand(s) up and down or back and forth" rhythmically in time with speech (McNeill 2005, 40). They are, McNeill (2005) said, "among the least elaborate of gestures" (40). Beat gesticulations differ from the other three types of gesticulations in that they do not refer to some entity, whether present or absent or whether concrete or abstract. Rather, they serve as a metronome to speech, and as such they can signal items of importance, such as important spoken words. In fact, McNeill (2005) said that beat gesticulations are "the equivalent to using a yellow highlighter on a written text" (41). Figure 2.9 shows Tonya performing a beat gesticulation that emphasized the words "sense" and "applying" in her explanation for her suggestion that Stewart reread the chapters in the book that cover oxyacetylene welding: "Because they'll make a lot more sense now that you're actually applying it."

These four types of gesticulations accompanied teachers' and students' verbal communication and enriched teachers' scaffolding of embodied knowledge, as when Tom used an iconic gesticulation to help convey brushing silica off a root weld (see excerpt I.2).

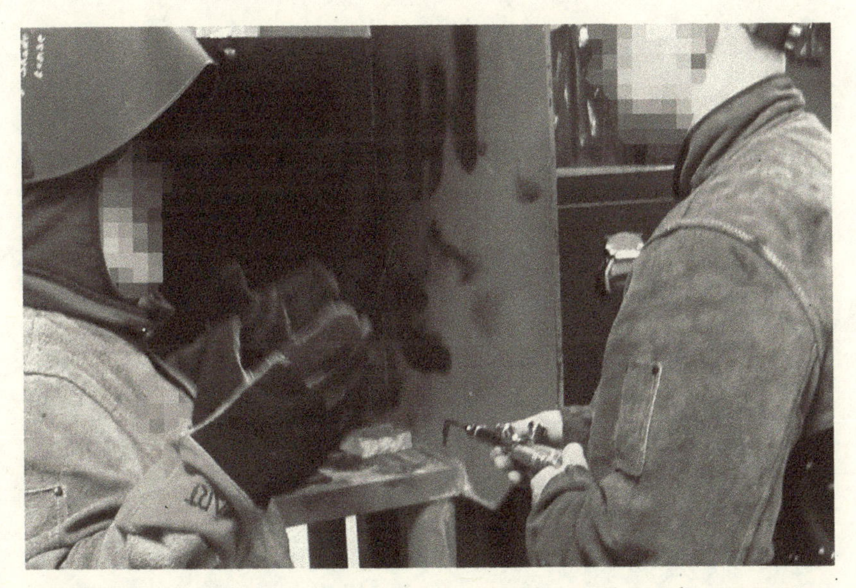

Figure 2.9. Tonya used a beat gesticulation in rhythm to the explaining strategy, "Because they'll make a lot more sense now that you're actually applying it," referring to the chapters that Stewart might reread.

Conclusion

In this chapter, I have tried to explain how my views about women in welding—both in industry and in education—have been complicated by my own experiences and by my conversations with women welding teachers and students. I have also explained my mixed-method approach to the study of teachers' technical communication during scaffolded teaching of embodied knowledge. In this book, via an autoethnographic approach, I report some of my own experiences as a learner in various welding classes. I also relate how, via a discourse analytic approach, I closely examined the ways that welding teachers used tutoring strategies and gesticulations to co-construct embodied knowledge with welding students. In the next chapter, I begin a three-chapter examination of scaffolded teaching. I start in chapter 3, examining how teachers used instruction strategies combined with gesticulations to scaffold students' learning of equipment setup, body position, and technique.

Chapter 3

Teaching Core Components of Welding

Scaffolded teaching is contingent; teachers base their interventions on what students have said or done in response to teachers' diagnosis of students' current level of understanding. This chapter examines how the welding teachers used instructional strategies (and accompanying gesticulations) as their contingent response while teaching three critical elements of producing a good weld: equipment setup, body position, and technique. In scaffolded teaching, teachers use instruction when they determine that a student's current level of understanding warrants greater direction and fewer "degrees of freedom" on the path to the desired outcome (Wood, Bruner, and Ross 1976, 98). With instruction, teachers supply solutions and explanations. Table 3.1 lists the instruction strategies that this study identified and briefly defines them.

Table 3.1. The instruction strategies found in this study's data

Strategy	**Definition**
Telling	Teacher directs the student in what to do, using little or no mitigation to lower the face threat of advice.
Suggesting	Teacher directs the student in what to do, using more mitigation (often negative politeness) to lower the face threat of advice.
Describing	Teacher relates the characteristics of a thing or action, sometimes with metaphor.
Explaining	Teacher offers reasons for a given assertion or directive.
Demonstrating*	Teacher shows how to perform a task or position.

*A nonverbal tutoring strategy.

The Role of Instruction in Teaching Equipment Setup

One of the most challenging tasks for a welding student—at least a new welding student—is getting set up to weld. This setup might involve hooking up and turning on gas, attaching needed machine accessories, such as a stinger[1] for SMAW, and choosing correct options on the machine for the process, such as switching the polarity of the welding machine from positive to negative. Forgetting setup tasks like these will produce very ugly welds or make it so that the machine doesn't work at all. My habit, even after a year of welding school, was to forget to turn on the gas. The result of this error is porosity—an easily identifiable defect characterized by multiple small holes, or pores, in the weld. Getting set up, for me, could take as long as a half hour—more if I needed to prep my base metal by grinding down its surface first.

After ensuring that the machine is ready to run, a student must determine the appropriate settings for a given process, position, and material. For example, in GMAW, a welder would use a different voltage welding a lap joint in the horizontal position than in the vertical-up position and would turn the voltage down for thinner base metal. In my own classes and in some of the classes that I observed, students sometimes watched videos produced by the Hobart Welding Institute of Technology to see a particular joint and position carried out, and these videos provided the range of settings that students should use when trying the weld themselves. Even after several semesters in welding school, machine settings remained somewhat mysterious to me. I didn't exactly understand *why* I needed any particular setting.

To keep up with my classmates and to save time, I wrote down the settings that I needed for each process and position in a little notebook that I kept in my tool bag. I got the idea for this notebook from one of my welding buddies, Serge. I met Serge in my first semester at DMACC. We were in the lab at the same time for three semesters, although he was ahead of me in the program and thus was enrolled in different classes. Serge was one of the best students in the program (though he'd probably humbly deny this). He kept a little notebook, on the cover of which he'd written "Important Shit." I decided that I too needed a little notebook in my bag to keep track of my important shit. So, in short, I wasn't clear on why I needed a particular setting for a given process, metal thickness, and joint. I just wrote down each combination of numbers in my notebook and used the same ones the next time.

Teachers wanted students to develop a deep understanding of the abstract concepts that govern welding machine settings, such as use of positive or negative polarity and the relationship between amperage and wire speed. But they recognized that such understanding takes time. And in the meantime, they wanted students to get experience with using the machine and determining for themselves the perfect setting for a particular weld joint and position on a particular machine. While teaching machine setup then, teachers diagnosed students' level of understanding, determining whether students—like me—needed more direction or whether they could take some control over their own learning.

Such was the case when Tom, making his tour around the lab to meet one-to-one with students, stopped at Sam's booth. Sam had been practicing pulse-spray transfer (a type of GMAW that cycles between a high and low current) in the vertical-up position. At the moment of Tom's arrival, Sam offered up an assessment of the quality of his work, as if to anticipate a potential negative evaluation from Tom, saying that his first pass at a vertical-up weld had been "terrible." In response, Tom looked over the welds that Sam had created thus far and diagnosed a potential problem that might have generated Sam's results—Sam's technique. Based on this diagnosis, Tom moved into position to demonstrate the proper technique, a wider weave pattern. In the process, Tom diagnosed a problem in the machine settings that Sam had been using:

Excerpt 3.1

SAM: Even the first pass was all- It was terrible.

TOM: Ok. Yeah I have a feeling it's we need to uh- we need to widen out our weave just a little bit. [Sam preps the T-joint, 20s]

SAM: You want to turn on the [unclear]?

TOM: Sure. We'll start on your first set. [Tom welds, 10s]

After trying for 10 seconds to weld with the settings that Sam had been using, Tom stopped to make adjustments, uttering a directive to himself, which served as a describing strategy: "Turn that arc length up just a little bit." He moved outside the booth to adjust the machine settings, raising "arc length" and lowering the wire speed. Afterward, he used another describing strategy to reinforce his demonstration ("So I'm turning the arc length up and the wire speed down") and an explaining strategy ("So that we try to get more of a spray but a little less wire coming out") to clarify for Sam the effects these adjustments would have on the weld:

Excerpt 3.1 (continued)

TOM: Turn that arc length up just a little bit. [Tom adjusts the machine's settings] So I'm turning the arc length up and the wire speed down. So that we try to get more of a spray but a little bit less wire coming out.

SAM: Ok. [Tom welds again, 36s]

After 36 seconds of welding to test the new settings, Tom described the new results he had obtained from the readjusted settings: "So it's starting to flatten out." The describing strategy that Tom used illustrates how welding teachers' describing strategies resembled think-aloud protocols common in user-experience research, where researchers ask study participants to articulate what they are thinking as they try to perform some task with a website or app (see, e.g., Fan, Shi, and Truong 2020). As they perform the task, participants state out loud what they see and hear, much like when Tom described his own actions during his demonstration, and then state their conclusions about how to perform the task, just as Tom explained the reasons for actions that he had just performed and described.

The utility of teachers' thinking out loud to describe what they perceive as well as to explain their reasoning has been noted in other pedagogical settings involving embodied knowledge, particularly in research related to surgical instruction. For example, in their study of surgical-resident training, Ruoranen, Antikainen, and Eteläpelto (2017) noted that "[i]t is especially important for trainers to make their own reasoning transparent, to think aloud" (329; see also, Sutkin, Littleton, and Kanter 2015a,b). Like the welding teachers in this study, the surgical trainers reinforced their demonstrating—nonverbal input—with verbal descriptions and explanations, and thus created a multimodal instructional intervention.

The ability to describe and explain while demonstrating does not necessarily come naturally. When I asked Tom whether welding while teaching is difficult, he answered with a strong affirmative:

TOM: Absolutely. That took me almost a year in class to really start to develop a- a- just a way of handling that, um. When I- when I first started every time I would start to talk my weld would not turn out well. Um, so it took me a little bit to, you know, kind of get my brain wired to the point where I could talk about it while welding.

Further, when I asked him about the reason that he had decided to practice this pedagogical skill, he responded:

TOM: I had interactions with students where I would run a weld, stop, and then try and talk about it and oftentimes they came away almost more confused than if I hadn't done anything.

Like the surgical trainers of the Ruoranen, Antikainen, and Eteläpelto (2017) study, welding teachers have had to find ways to layer skillful demonstration with effective teaching. In my mind, this critical ability is one that shouldn't be left to teachers of embodied knowledge to figure out for themselves. This skill is one that should be taught along with other pedagogical practices in teacher training.

Tom then moved from the instruction of describing to the cognitive scaffolding of a pumping question: "So you can see how dramatic that is?" With this question, Tom assessed intersubjectivity—that characteristic of scaffolded teaching that checks the extent to which a shared understanding of the task exists. More specifically, Tom's question seemed aimed at determining the extent to which Sam perceived what Tom was trying to help him to see: the difference between the results Sam had gotten with the previous settings and the results that Tom was getting with the adjusted settings. Tom's pumping question had the added benefit of generating interactivity, another critical component of scaffolded teaching. With this question, Tom brought Sam into the conversation:

Excerpt 3.1 (continued)

TOM: So it's starting to flatten out. So you can see how dramatic that is?

Sam: [unclear] try a little.

TOM: I was- I was just going to try that real quick and I don't think that works here.

Sam: [Ok.

TOM: [The puddle is too big and too fluid so it doesn't stack on top of itself very well. [Tom welds again, 30s] But it's a lot bigger. [2s]

In response to Tom's question, Sam asked Tom about some other potential option for getting a flatter weld, a question lost in the noise of the welding lab. But Tom's response to Sam's question was clear: "I was just going to try that real quick and I don't think that works here." With this response,

Tom offered Sam an explanation for why the other option that Sam had mentioned would likely not work—the puddle with pulse-spray transfer is too big and fluid and would not sufficiently layer on itself. Sam's question in response to Tom's pumping question exemplified the interactivity that even simple questions of perception, such as Tom's "Can you see?" question, could invite students into the conversation.

Seeming to sense that he had not yet sufficiently helped Sam perceive the result they needed—a larger, more fluid puddle, but one not so big that it doesn't stack—Tom continued first with another type of cognitive scaffolding strategy, referring to a previous topic ("your technique is similar to short circuit"). With this strategy, teachers juxtaposed new information, in this case, pulse-spray transfer, against known information, in this case, the more familiar short-circuit welding. Referring to a previous topic showcased how teachers worked within students' ZPD, building stepwise on what students already knew. And, it renegotiated intersubjectivity in that it attempted to reshape the shared understanding of the task. After, Tom moved back to instruction, reinforcing with another describing strategy the desired appearance of the puddle: "the puddle is much bigger and much more fluid":

Excerpt 3.1 (continued)

TOM: Well one of the- one of the things with this when you get out of position, your technique is similar to short circuit but you have to remember the puddle is much bigger and much more fluid. [Hands gun back to Sam, 2s]

With this describing strategy, Tom helped Sam identify the characteristics of the puddle that he should try to achieve in this process and (out-of-position) position.

And this time, Tom layered an iconic gesticulation onto his describing strategy, illustrating the result—a bigger and more fluid puddle—that they had just achieved with the new settings. He repeated an iconic gesticulation, performing the nose-to-chest movement once when he said "is much bigger" (figures 3.1 and 3.2) and repeating the gesticulation again when he said "much more fluid." The duplicated gesture, starting high (in front of his face) and finishing low (at his chest), indicated the puddle's increased size and its new ability to spread out.

This interaction between Tom and Sam reveals how demonstrating, combined with describing and explaining strategies, can serve as an

Figure 3.1. Tom started his iconic gesticulation aimed at illustrating the size of the puddle as he said "The puddle is much bigger."

Figure 3.2. Tom finished his iconic gesticulation aimed at illustrating the size of the puddle as he said "The puddle is much bigger."

effective intervention for students who struggle with machine setup, one that balances students' need to use class time to practice weld technique and students' need to understand the task's desired outcome (a fluid, flat puddle) and the machine settings that would generate that outcome. With this strategy cluster, teachers contingently responded with instruction to students' current level of understanding, giving them direction to help them tie machine settings to puddle outcomes and at the same time move along toward practicing more welds with the proper technique.

Tom finished this demonstration by telling Sam he could "play" with the settings in order to generate a range of different results. My own teachers said the same to me a few times. They invited me to "play around" with the settings to see for myself the different outcomes of, for example, higher or lower voltage. When a teacher assesses that a student understands the desired outcome and the path toward that outcome, a teacher might invite the student to achieve further understanding, in this case, by purposefully generating bad welds. Personally, playing around with settings is something that I never wanted to do once I had found the machine settings that seemed to work. Rather, I wrote down the settings that worked in my Important Shit notebook and used my lab time to practice laying the best welds that I could. Students with more confidence—students like Samantha—are much more likely, it seems to me, to test the limits of acceptable settings. For me, purposefully creating bad welds held little allure. I accomplished bad welds without trying.

Tom had diagnosed Sam's level of understanding based on Sam's weld results and had used demonstration, describing, and explaining as a clustered intervention. But when a teacher perceived a student to have a somewhat higher level of understanding, the delivery of the intervention could shift contingent upon that level of understanding. Seth was a more-advanced welding student than Sam, needing just two GTAW classes to complete in his program, and Teresa's interaction with him acknowledged his growing expertise.

Just before the talk in excerpt 3.2, Teresa had come to Seth's booth to help Seth after he had reported to her that his welds looked bad, outcomes that he thought stemmed from a gas problem. In response, Teresa left Seth's booth, opened a valve on a bank of gas canisters located outside the lab, and returned to Seth's booth to run a bead to check the weld quality. After running a weld and seeing that the problem was solved, Teresa assured Seth that he should be "all set." She did, however, notice at that moment the settings on his machine. Diagnosing his level of understanding based on these settings, Teresa took the opportunity to advise Seth on machine setup in light of his upcoming shift from the horizontal to the more difficult vertical-down position:

Excerpt 3.2

TERESA: Um you might notice because you were running um flat with the last one you had to do- or was it [horizontal?
SETH: [It was horizontal.
TERESA: Horizontal? With the vertical down you might notice you can run it hotter and a higher wire speed.
SETH: Yeah. With a higher wire speed?
TERESA: Yep.
SETH: Yeah. I remember last time it was at the lower end.
TERESA: Ok. Yeah. Try it a little higher I mean- Yeah. 170 is a good place to start. 175.

Teresa used a suggesting strategy to point Seth toward higher settings for the vertical-down position: "you might notice you can run it hotter and a higher wire speed." Rather than more directly telling Seth to turn up the wire speed, for example, by saying "Run it hotter and with a higher wire speed," Teresa presented her advice as a hypothetical, a situation that might or might not occur and that he might or might not notice. In addition, her suggestion invoked Seth's ability to perceive expertly, to adjust his machine settings based on his weld outcomes. In other words, even as she conveyed instruction, Teresa acknowledged Seth's growing knowledge.

Seemingly a bit surprised by Teresa's suggestion, Seth asked for confirmation that Teresa had actually meant "higher," particularly in light of his prior experience: "I remember last time it was at the lower end." Seth's confirmation question generated interactivity, the critical component of scaffolded teaching and learning that allows students to adjust their current understanding. Such interactivity more effectively generates co-constructed knowledge.

In addition, according to van de Pol, Volman, and Beishuizen (2012), such interactivity is even more effective when it manifests openness, that is, when both the student and the teacher are open to the ideas of the other (201). Though Teresa was certain of the efficacy of her suggestion, her response indicated an openness to Seth's doubt. First, though, she shifted from suggesting to telling, more forcefully stating her advice: "Try it a little higher." But then, she acknowledged Seth's assessment of the situation, saying that his current setting, 170, was "a good place to start." Critically though, she offhandedly tacked on the possibility of 175, a higher setting more in line with her own thinking, as yet another good starting point.

In this interaction, then, Teresa conveyed the instruction she thought Seth needed, but at the same time she acknowledged his welding experience and, as part of that, showed an openness to his ideas. In these ways, she responded contingently to Seth's level of understanding.

The Role of Instruction in Teaching Body Position

Welding teachers spent much of the class period moving from booth to booth, watching students as they welded or evaluating the welds that students had recently completed. Through such assessments and diagnoses, teachers might intervene with instruction. Above, I discussed how teachers' demonstrating worked with their describing and explaining as an instructional and thus directive intervention in students' progression toward expertise. Demonstration—creating an idealized performance of a task (Wood, Bruner, and Ross 1976, 98)—played a strong role as well in teacher' interventions in students' body positioning. My GMAW teacher, for example, came to my booth again and again to demonstrate how I should position my body for an overhead weld. (Figure 3.3 shows some of the burns I obtained as I practiced overhead welding.)

Teachers' demonstrations related to all aspects of welding, from changing out a spool of wire on a welding machine, to safely holding a piece of metal to the heavy-duty grinder to grind down a weld cross-section for a bend test, to using the belt grinder to get millscale off the base metal, to

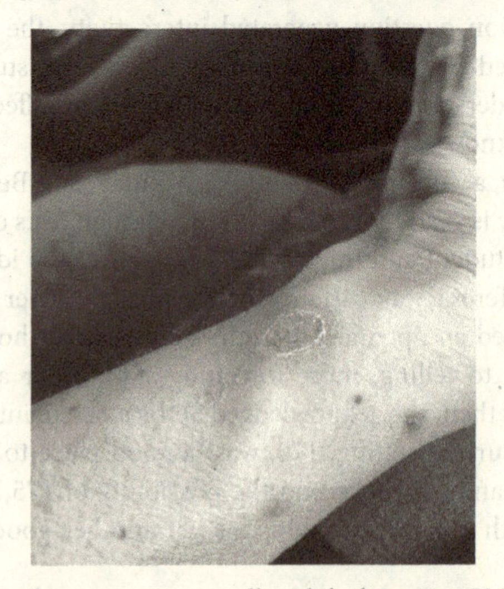

Figure 3.3. I burned my arm repeatedly while learning GMAW, SMAW, and FCAW in the overhead position.

striking an arc and laying a bead, as Tom did in excerpt 3.1. When I asked Tom whether demonstrating was common practice for him, he answered:

TOM: Absolutely. Absolutely. Um, because it's one of those things that- there- I mean there are definitely students that you can talk to them all day but until you can describe what's going on while it's happening it really doesn't sink in for them.

Indeed, all the teachers, including my own, consistently employed this strategy, especially in relation to body position. And they did so when they diagnosed students as needing more direction in how to situate their bodies at the weld table or how to position items in the booth for good ergonomics, for example, the best place to put the stool when sitting (for GTAW or oxyacetylene welding) or the best height and placement of the table's arm for the student's height and for the weld position.

For example, Ted visited Spencer's booth while making his rounds around the lab. After observing Spencer as he completed an overhead SMAW weld, Ted diagnosed what Spencer still needed to learn: the importance of positioning his body so that he could hold his arms steady for the entire length of the weld and use proper technique. Responding to this diagnosis, Ted first repositioned the table arm to better fit Spencer's height. Then, Ted layered instruction via suggesting and telling strategies onto his instruction via demonstration (figure 3.4):

Excerpt 3.3

TED: Here's what I'm going to recommend. Let's take this down [Lowers the arm on the welding table, 4 sec] Ok. I'm going to recommend that you get yourself- Step into this.

SPENCER: Yep.

TED: Ok? Step into this. [2s] And you, get this tight, to you. Ok?

Ted began his verbal instruction with a suggestion: "I'm going to recommend that you get yourself." Ted the switched to more directive telling strategies: "step into this" and "you get this tight to you." By layering verbal over visual instruction in his intervention, Ted enriched his input into Spencer's learning. Afterward, Ted offered further instruction, this time focused on Spencer's technique with the electrode. I discuss this part of the interaction later in this chapter when I discuss teachers' interventions in students' welding technique.

Tom visited Stephanie's booth not on his usual rounds around the lab but because, earlier, she had asked him to come over to help her with

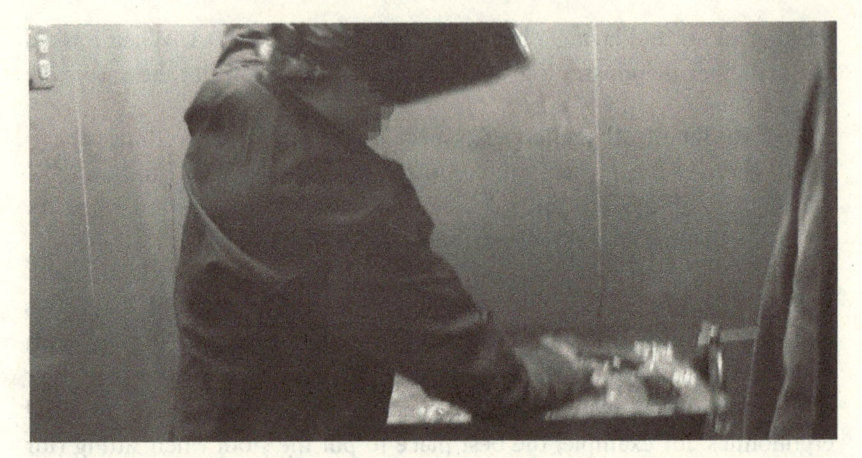

Figure 3.4. Ted told Spencer to "step into this" as he demonstrated the body position that would facilitate holding his arms in the correct position.

the task—an advanced one—of welding around a pipe. To diagnose the problem Stephanie had been having with the task, Tom urged Stephanie to perform a dry run of a pipe weld. In a dry run, welders ensure that they can perform a given weld effectively and, preferably, comfortably. They try to find a position that allows them to brace themselves against something solid (such as the weld table or a wall) so that they can stabilize themselves. Welding teachers commonly recommend dry runs of welds, even when the weld is just the length of a six-inch coupon. Doing a quick dry run even for a simple weld helps ensure correct work and travel angles[2] throughout the length of the weld, and it allows welders to practice the correct speed for the weld as well. Depending on the position of the weld, it can also help ensure that the welder's arm will move smoothly over a table or other surface used for stability.

Tom started the interaction by assessing the weld Stephanie had already completed, offering a bit of praise to reassure Stephanie at the outset ("I mean this is looking pretty good"). Then, he asked a series of questions that diagnosed Stephanie's current level of embodied knowledge, specifically, how Stephanie had thus far been positioning her body and the gun in relation to the pipe:

Excerpt 3.4

Tom:	I mean- I mean this is looking pretty good. You said you have trouble down on the bottom? [2s] How are you positioning when you do it?

STEPHANIE: Like on my [unclear]?
TOM: Yeah.
STEPHANIE: I'm not resting my arm on anything- I'm really just, coming around. [Is that a
TOM: [Ok. So are
STEPHANIE: problem?
TOM: you, like, here? And then you're squatting down?
STEPHANIE: Yeah.

By observing Stephanie's dry run, Tom sussed out the likely problem with Stephanie's body position and suggested an alternative:

Excerpt 3.4 (continued)
TOM: That's probably your- As you- What I would suggest is try to get set up in such a way that you can bend at the waist. You can start at the top and work your way to the bottom. So what that looks like for me- Here, let's swap spots. [2s]

The two swapped spots, and Tom then began a demonstration. He modeled a flowing body movement around the pipe, coupled with the angle and speed of the gun required to maintain a consistent weld. While he demonstrated, he used a string of describing strategies to reinforce the actions he was carrying out:

Excerpt 3.4 (continued)
TOM: I would start, here. And then it's just a bend at the waist. And my relation between the gun and my face really doesn't change. Ok? So, what it looks like is I'll start here. And it's a- It's slow. Start with the normal [unclear] so flat. Now we're going into vertical down. And now is when I start bending at the waist. [2s] Been awhile since I've had one of those. Spark flew right in my ear. Ok. So to continue.[3] [6s] Now I'm into overhead technique. [7s] So it's, just that method [2s] of starting where you're comfortable, and it's just a roooll around the pipe. Ok? So give that a shot. You might need to lower this [the table arm] is a little bit. I think you're up too high.
STEPHANIE: Ok.

Tom's describing strategies related to the position of his body ("And then it's just a bend at the waist") and the relation of his body to the weld surface ("And my relation between the gun and my face really doesn't change"). Tom also used this visual-verbal instruction as an opportunity for cognitive scaffolding, building stepwise on what Stephanie already knew—the weld positions of flat, vertical down, and overhead. In this way, slightly lessened his control over the interaction by pushing Stephanie to connect new information with known information.

One of Tom's descriptions of his movement around the pipe employed a rhetorical tool found in all the teachers' talk: metaphor. Metaphors allowed welding teachers to describe some less familiar target domain in terms of a more familiar source domain. Indeed, the word comes from Greek via Latin and French, meaning "to carry across." Metaphors fall into two main categories: conventional and novel. As mentioned in chapter 1, conventional metaphors—what Lakoff (2006) called everyday metaphors—operate unnoticed in language. They form our unconscious understanding of the world. Discussing Reddy's (1979) germinal article, "The Conduit Metaphor," Lakoff argued, "metaphor is a major and indispensable part of our ordinary, conventional way of conceptualizing the world" (186). Carter and Pitcher (2010) put it this way: A conventional metaphor is a conceptualization that has become successful in its invisibility (585). In welding, welds are conceptualized largely in terms of anatomy; they have a toe, a leg, a face, and a throat. They also have a root, which follows a different conceptual map. Other conventional metaphors in welding include keyhole, a hole at the edge of the weld puddle, and crater, the cavity at the end of a weld. These terms constitute common welding jargon (in English).

Novel metaphors, in contrast, are those created in the moment for a particular purpose. Because they are deliberate, "tuning devices," such as "sort of," "kind of," and "just," often (but not always) precede them (Cameron and Deignan 2003). For example, Scott stepped outside his booth to find Teresa and to ask her for an assessment of the pad of beads[4] he had run in the flat position. Using a pumping question to determine whether Scott perceived the same phenomenon that she did, Teresa associated Scott's pad of beads to rivers on the moon:

Excerpt 3.5

SCOTT: That's why-

TERESA: And that might be just because you're putting multiple beads on. The whole plate is getting hot. See how- You see these little rivers?

SCOTT: Yeah.

TERESA: Reminds me of the moon.[5]
SCOTT: Yeah.
TERESA: Kind of dry rivers. Versus like- This one's a little shinier.
SCOTT: Yeah.
TERESA: Um that's that heat difference.

Teresa signaled her conceptualization of Scott's results with the tuning device "kind of," signaling that Scott should not understand her description literally. The tuning device also suggested that she had generated this source-to-target mapping herself in an attempt to describe in the moment what she saw.

Similarly generating a novel metaphor to convey what she saw, Tonya assessed Shawn's work after observing him weld a vertical-up joint in GMAW. She too, used a tuning device before her metaphor, the downgrader "just." With it, she mitigated her metaphor's strength, suggesting that Shawn's weld was not too terribly flawed:

Excerpt 3.6
TONYA: Just get it to the side. Get it to the side. Get it to the side.
SHAWN: And plus I was bending down too.
TONYA: Well yeah. It's a lot more comfortable up here but it looks- it's a nice-looking bead. It's just got a little beer belly on there.
SHAWN: Did you say beer belly?
TONYA: [Laughs] Yeah. Well I don't know how else to describe it.

Tonya's metaphor was so novel that it made Shawn question whether he had heard her correctly. Whether conventional like "toe" or novel like "beer belly," metaphors tend to operate on a lexical level, where a source domain serves as a way to understand a target domain.

Returning to excerpt 3.4, what was so interesting about Tom's description, "it's just a roooll around the pipe," is that it created a novel metaphor on the phonetic level, with an elongation of the vowel in "roll," rather than on the lexical level. With this elongated vowel, Tom extended the word in time to reinforce the slowness and smoothness of the movement he had demonstrated and that Stephanie was trying to learn (figures 3.5 and 3.6). The elongated vowel generated a phonetic metaphor that helped Tom instruct Stephanie in proper body position and movement around the pipe.

Tom's phonetic metaphor was not the only one in this study's data. Another appeared when Teresa used a describing strategy to reflect back

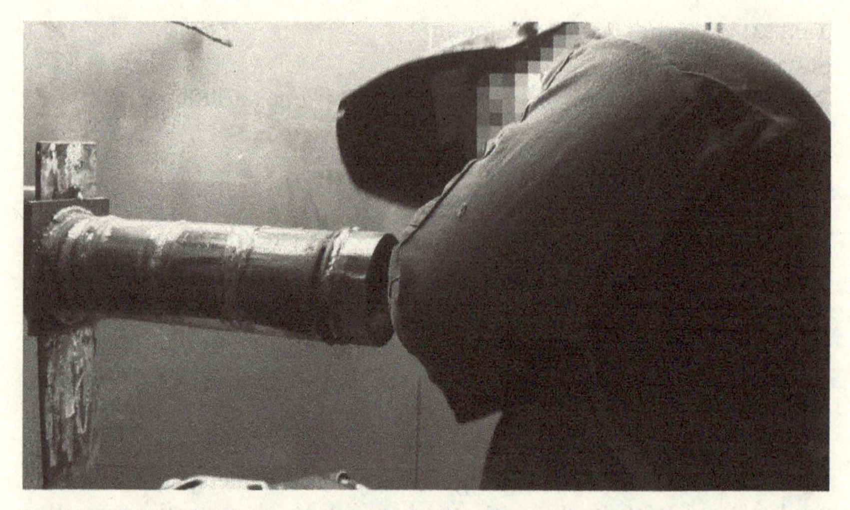

Figure 3.5. Tom began demonstrating body position for welding around a pipe while using a phonetic metaphor: "And it's just a roooll around the pipe."

Figure 3.6. Tom finished demonstrating body position for welding around a pipe while using a phonetic metaphor: "And it's just a roooll around the pipe."

on her just-completed demonstration. Talking to Sophia about a horizontal open-root joint, Teresa elongated the vowel in "all" to convey the movement of the electrode over the entire joint, including the tack at the end:

Excerpt 3.7

TERESA: Basically just, digging out between the plates so I can continue my weld. [3s] I'm coming back and I'm pushing through.

[Completes weld, 15s] And run it aaall the way to the end of the plate. Don't stop at the tack. [Run it all the way through

SOPHIA: [Ok.

TERESA: until the end of the plate.

Like Tom, Teresa tapped into paralinguistics to create a novel metaphor that mapped a phonetic source to a target movement in order to clarify her instruction.

The Role of Instruction in Teaching Technique

As teachers moved from booth to booth, they encountered opportunities to diagnose problems with students' techniques and to determine an appropriate, contingent intervention. As I have pointed out throughout this chapter so far, these interventions often layered verbal instruction strategies with demonstration, as seen in excerpts such as 3.1, 3.3, and 3.4. But scaffolded teaching with demonstration could employ silence as well, as when Tonya came to Stewart's booth to introduce him to oxy welding a lap joint in the flat position. Tonya was silent for seven seconds between her introduction to the demonstration and her first telling and explanation strategies:

Excerpt 3.8

TONYA: Now I'm going to do it hot. [7s] Don't be skimpy on that lower plate because it takes longer to get hot.

STEWART: Ok.

TONYA: [unclear] because that upward plate is going to get hot real easy. [1m, 59s]

After welding for nearly two minutes, Tonya finally spoke again; this time she used a describing strategy to highlight the characteristics that she was looking for in the weld puddle (having it spread across both coupons) and noted the result—that Stewart would be able to move the puddle:

Excerpt 3.8 (continued)

TONYA: When you've got that puddle on both then you can move it around.

STEWART: Ok.

Tonya's demonstration took over 13 minutes to complete and was the longest in my data. It showcased how teachers could engage in extended silence that seemed intended to allow students to concentrate on teachers' nonverbal instruction and the phenomena that they were seeing, hearing, and possibly even smelling.

The welding lab offered opportunities to learn techniques other than those directly applicable to running a bead, such as work angle, travel angle, travel speed, or weave pattern. Besides demonstrating welds, teachers demonstrated techniques for manipulating tools and electrodes. For example, Teresa responded to Suzanne's observation that feeding the filler-metal rod through her hand smoothly and consistently was quite difficult. GTAW welders like Suzanne must be able to feed the filler rod through their fingers without relying on their other hand, which is occupied with the tungsten electrode. Teresa showed Suzanne how to feed the rod through her fingers to maintain a supply of filler to the weld puddle while maintaining a safe distance between the puddle and her fingers. She demonstrated the movement that she would use if she were actually welding to show Suzanne how she might practice the skill (figure 3.7):

Excerpt 3.9

TERESA: There's a couple ways to do it, and you can take a little practice this- If you're kind of welding up to your fingers,

Figure 3.7. Teresa suggested that Suzanne practice "feeding it [the filler rod] with your thumb like this."

	you can go ahead and start practicing with feeding it with your thumb like this.
SUZANNE:	Ok.

Similarly, after demonstrating a body position that would help Spencer perform an overhead weld (see excerpt 3.3 and figure 3.4), Ted demonstrated a "pool-cueing" technique for sliding the SMAW electrode into the weld puddle (figure 3.8):

Excerpt 3.10

TED:	And then start working it like this. Gradually pushing it up. Ok?
SPENCER:	Yeah.

When I asked him about this term, what I first perceived to be a novel metaphor, Ted told me that the descriptive name is well-known term in pipe welding used to denote a common technique. It seems, then, that pool-cueing is another example of conventional metaphor in welding. It has become a commonplace way for welders to talk about this technique; that is to say, it's become jargon.

Excerpts such as 3.8, 3.9, and 3.10 show how teachers demonstrated technique by using items in the immediate environment to intervene in students' learning, notably the welding gun or torch but also filler-metal

Figure 3.8. Ted told Spencer, "And then start working it like this. Gradually pushing it up. Ok?" as he demonstrated the technique of pool-cueing the electrode.

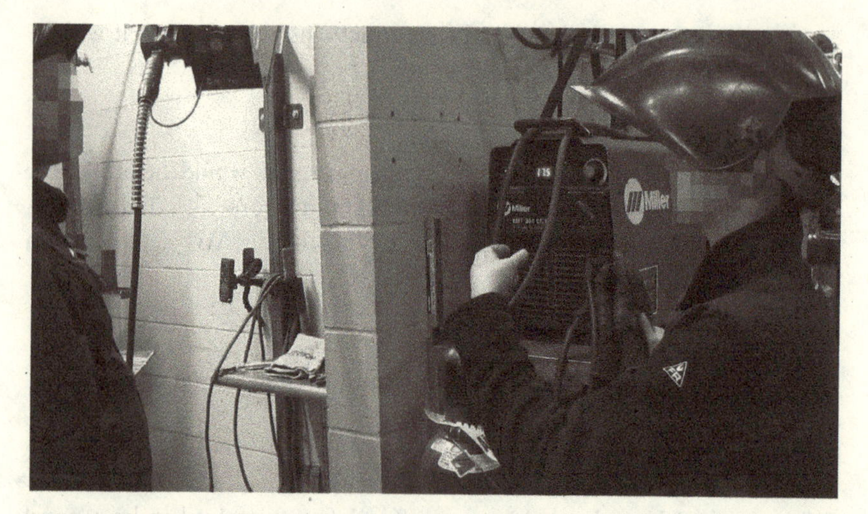

Figure 3.9. Teresa told Seth, "[B]ut really think about *push*ing through to the other side."

Figure 3.10. As Seth said, "on the leading edge," he made an abstract deictic gesture when he pointed to the leading edge of a future puddle.

rods and electrodes. But when they didn't have quick access to such items, they relied on gesticulations, particularly iconic gesticulations. After Teresa helped Seth with the setup of his welding machine (see excerpt 3.2), she instructed him on a how to proceed with the more advanced position of vertical down. Teresa began with a telling strategy ("really think about *push*ing through to the other side"), which she reinforced with an iconic gesture (figure 3.9).

Excerpt 3.11

TERESA: And then if when you're dragging it it's a slight drag but really think about *push*ing through to the other side.

As she spoke, emphasizing the syllable "push," Teresa mimicked the movement of the electrode with her finger. As noted above, Seth was an advanced student and already employed as a welder, and his advanced status likely played a role in the interactivity that he brought to the interaction. Seth built on Teresa's instruction, noting yet another concern about technique that he had, one based on his previous attempts in horizontal position:

Excerpt 3.11 (continued)

SETH: I just need to make sure I can stay on the leading edge of that puddle. That- that was my issue.

Seth's response added to what Teresa had advised and, more important to scaffolded teaching and learning, showed his ability to assess his own past performance—an important step on the path to mastery. In responding, Seth employed an abstract deictic gesticulation (figure 3.10), pointing to a nonexistent object—the leading edge of some near-future weld. With this gesture, Seth showed his understanding the electrode's proper location in relation to the puddle and, in the process, created further intersubjectivity between himself and Teresa.

Indeed, Seth continued his prescription for success, even overlapping with Teresa's:

Excerpt 3.11 (continued)

TERESA: Getting in [unclear] of you a little bit. Yeah.
[Well why don't-

SETH: [Or I'm too far down.

TERESA: Ok. Run a couple and see how it goes.

Seth's meta-awareness of his learning seemed to give Teresa confidence that she could safely fade and let him "run a couple" vertical-down welds on his own without further instruction.

So far, I have discussed how teachers intervened in students' learning with instruction, directing students in what they should do. I have focused in part on teachers' demonstrations, those idealized performances for imitation. But welding teachers also demonstrated techniques that students should avoid. Asplund and Kilbrink (2018) described similar demonstrations in their case study of a welding teacher and a plumbing student in a Swedish upper-secondary technical school. The student was learning backhand welding, which means dragging the puddle instead of pushing it. This technique is common in welding processes that produce slag, such as SMAW.[6] They noted that the teacher made the angle of the gun salient by contrasting correct and incorrect gun angles, telling the student:

> "The important thing is . . . that when you weld like this that you keep 90 degrees like this. . . . That is, not like this." (8)

The teachers in this study used demonstration to contrast effective and ineffective technique as well. For example, after Teresa told Sophia how to weld a horizontal open-root joint (see excerpts I.3, 3.12, and 4.6), she contrasted those efficacious steps against common mistakes. One mistake, she warned, was overheating the base metal. The other, she said, involved pulling the electrode out of the joint, an error that she acted out with an electrode (figure 3.11). Then, Teresa used a metaphor in describing the result of pulling the electrode out of the weld joint—an "umbrella of heat"—and used a metaphoric gesticulation to help Sophia visualize the problem that would arise (figure 3.12):

Excerpt 3.12

TERESA: Then the other mistake a lot of people make is pulling the rod out of the plates. Once you're starting your weld and your plate is hot- You're performing this weld completely between the two plates. You don't want to pull it out of the plates. It gives it kind of an umbrella of heat, and it will give you too big of a keyhole. It will overheat it-

SOPHIA: Mmhm.

TERESA: And you'll be blowing holes in it as well.

SOPHIA: Ok.

With her negative demonstration, and its associated telling, describing, and explaining strategies, Teresa made salient the differences between the

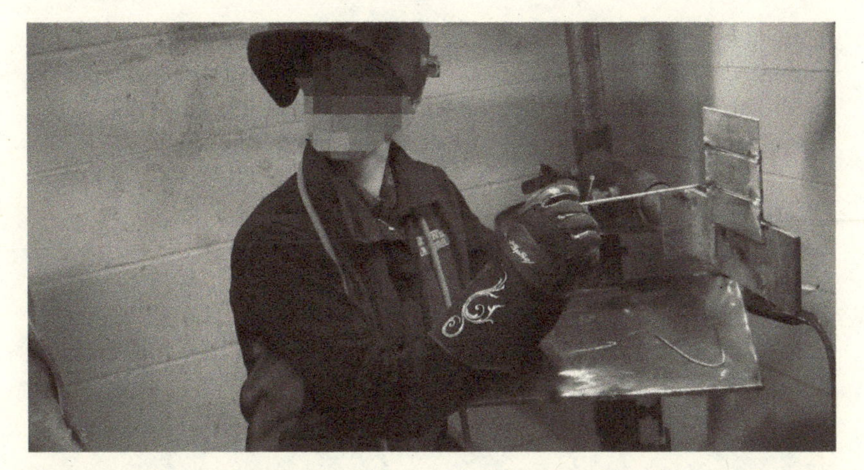

Figure 3.11. Teresa demonstrated the unwanted action of pulling the electrode out of the weld joint as she directed, "You don't want to pull it out of the plates."

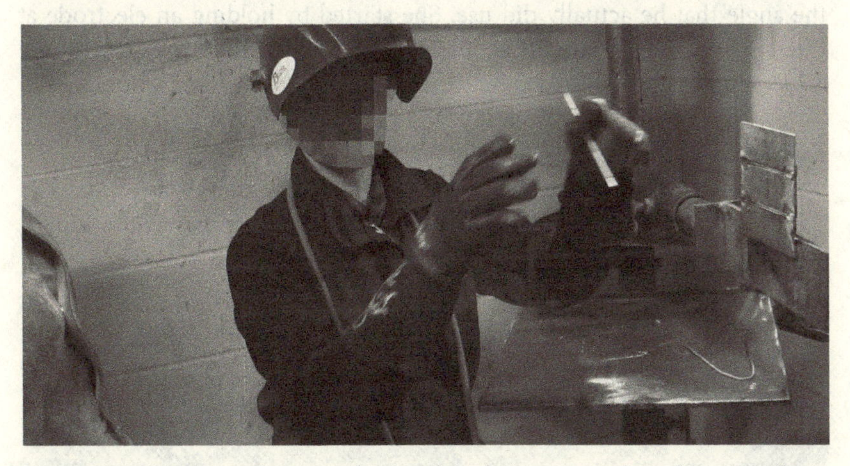

Figure 3.12. Teresa used a metaphoric gesticulation to reinforce her metaphoric description: "It gives it kind of an umbrella of heat."

correct actions and the incorrect actions that many newcomers make as they attempt this challenging weld for the first time.[7] Like instruction sets that tell users which step to take before pointing out which action to avoid, Teresa intervened with thorough instruction that Sophia, as a novice, needed to begin her time on this task.

Another way that teachers used this intervention of negative demonstration was to reenact what they had just observed students do. For example, after watching Simon complete a vertical-up weld in SMAW (offering moment-to-moment instruction via telling and explaining strategies throughout), Tonya contrasted what Simon should do against what she had just seen him do. She began by articulating the problem that she had just observed, stating it as an assumption of a problem that Simon himself noticed:

Excerpt 3.13

TONYA: The only thing that I would do different-

SIMON: What?

TONYA: Is the angle on the electrode. When you first start- Because I- When you get to the top and you noticed that you're having a lot of trouble controlling the puddle-

SIMON: Mmhm.

After diagnosing the problem, Tonya intervened with demonstration. Specifically, she contrasted the angle that Simon should have used against the angle that he actually did use. She started by holding an electrode at the correct angle (figure 3.13), and then pointed it slightly down (figure 3.14) to show the difference between effective and ineffective:

Excerpt 3.13 (continued)

TONYA: Because it's pointing down a little bit. We don't want it to point down. We want to be- By the time you get to the top you should be perpendicular to the plate.

SIMON: Ok.

Figure 3.13. Tonya demonstrated the difference between an effective and an ineffective travel angle with an electrode. She started the demonstration with the electrode pointed slightly up.

Figure 3.14. Tonya then pivoted her forearm at the elbow to point the tip of the electrode downward.

Tonya used demonstration, then, to contrast the electrode angle that Simon should imitate against a sort of live replay of the misstep in electrode angle that had detracted from his success. With this contrast in mind, Simon was primed for another attempt.

Conclusion

One critical characteristic of scaffolded teaching is teachers' diagnosis—and ongoing diagnosis—of students' current level of understanding. This chapter has examined interactions that occurred when teachers determined that students' learning required instruction, directive intervention often comprised of demonstrations overlaid with telling, suggestion, describing, and explaining strategies. Diagnoses occurred in several ways. Sometimes, teachers knew that students were completely new to the task at hand. Tonya knew that Stewart had no experience in oxyacetylene welding, so her intervention—a demonstration with telling and explaining strategies—was instruction (see excerpt 3.8). Similarly, Teresa knew that Sophia had never welded an open-root joint, so she too offered a demonstration (see excerpt 3.7). As noted in this book's introduction, these demonstrations were planned; sometimes, when teachers knew a student was ready for a new task, they demonstrated it when they arrived at the students' booth.

On a few occasions, teachers gathered several students to a booth to watch a demonstration. During these demonstrations, teachers used verbal strategies. Telling and suggesting strategies directed students in what to do.

In situations where students were novices to the task, instruction certainly seemed the effective choice. As McLain (2018) wrote in his study of design and technology teachers' demonstrations, "when teaching a new concept or skill . . . the choice might be to adopt a more restrictive and teacher-led approach," which "will limit the range (and potentially the creativity) of outcomes" (988). In contrast, he wrote, "a more expansive approach (where learner[s] potentially make more choices) can result in a broader range of outcome[s], which might be less skillfully realised if the requisite skills have not already been developed" (988). In the interactions like the one between Teresa and Sophia, students were complete newcomers to the task they were learning and thus needed more support.

In other cases, upon arriving at a student's booth, diagnosing the student's current level of understanding was somewhat more complicated. Teachers sometimes asked students to weld so that they could observe them and better diagnose problems with their body position or technique, as Tonya did when she visited Simon and Shawn (see excerpts 3.6 and 3.13) and as Ted did when he visited Spencer (see excerpt 3.3). Sometimes, during such observations, teachers issued telling and suggesting strategies in the moment. Tonya likened such observation and concurrent instruction to coaching:

TONYA: I feel like a coach. Like I'm coaching. Especially with the one-on-one when you're watching someone weld. And I'll say, "Get closer. Go faster. Slow down." You know. Try to coach them to figure out what they're doing wrong so they can do it right.

Observation was teachers' means of obtaining direct evidence of students' level of understanding—their level of embodied knowledge. One question that arose upon analysis of teachers' observations is whether teachers should, as Tonya indicated, provide instruction while students are performing a task or whether they should wait until students have finished. That is, to what extent does correction in the moment transfer to subsequent attempts, enabling students to self-correct? A longitudinal study of teaching and learning a skilled trade would help answer this question.

More often, diagnosis occurred when teachers visited students' booths and evaluated students' work thus far, typically the welds that they had produced, as when Tom visited Sam's booth. Tom noted that the problems both saw in Sam's welds were likely generated not only by Sam's technique, but also by faulty machine settings. Tom responded to Sam with instruction—demonstration first of adjusting the machine and then of the proper technique (see excerpt 3.1). Other times, as in the case of Tom and Stephanie, students approached teachers with questions (see excerpt 3.4). In such cases, students' interactivity—their questions—helped to clarify their level of understanding. Teachers responded with instruction, meeting students at their current level of understanding.

In some of these interactions, particularly ones in which students were not complete newcomers to the task, teachers might have attempted less directive interventions first, moving to instruction afterward if the level of understanding conveyed in students' responses warranted fewer degrees of freedom (Cromley and Azevedo 2005, 87). For example, they might have used cognitive scaffolding strategies, which are less directive than instruction. Pumping questions such as Tom's "So you can see how dramatic that is?" generated greater interactivity—even if only through the student's yes-or-no response. Pumping questions—even yes-no questions—push students to think for themselves and to contribute to the interaction. Open-ended pumping questions, such as asking students to describe the results of changes to machine settings, body position, or technique, can push students' thinking even further. Asking such questions can help students think about the reasons behind variables of machine setup, body position, and techniques. For example, they can help get welding students to think out loud about why a certain technique would work best for a given process and position and thus would push students to expend some cognitive effort on connecting actions to the reasons behind those actions. If a student isn't able to produce a response, the teacher can reduce students' degrees of freedom in responding by falling back to instruction and giving more support to keep the student on track to learning how to perform the task.

But as mentioned earlier, lab time is finite. So particularly in relation to machine setup, teachers must consider how much time they can allow students to set up their machines—a skill that students must learn—and how much time students will have to practice welding. Tom seemed to strike an effective balance between these competing needs. He demonstrated machine setup, describing the settings as he did and explaining his choices. Rather than preemptively set up the machine for Sam, he tied the settings to the results that he was getting. The demonstration combined with describing and explaining connected lab practice (procedural knowledge such as grounding the welding machine) to classroom learning (declarative knowledge such as direct electrical current). Such multimodal demonstrations to tie what the teacher is doing to why the teacher is doing it seem ideal.

All of the teachers talked as they demonstrated, though as Tom pointed out, learning how to carry out a complex task like performing a weld (particularly in a difficult weld position), and simultaneously thinking aloud about the process, took practice. Tom's comment raises questions about the way in which welding teachers and others who teach skilled trades learn pedagogical best practices in their field. What resources are available to help teachers learn how to, for example, demonstrate a task and describe and explain it at the same time? Research on best practices of teaching skilled trades is scarce. Few academic journals address the needs of skilled-trades teachers; the *Journal of Vocational Education and Training*, a journal with truly international scope, is one notable exception.

Although using both the nonverbal strategy of demonstration simultaneously with the verbal strategies of describing and explaining (along with other strategies) raises teachers' cognitive load, it produces dual-mode input—the verbal combined with the nonverbal—that likely benefits students' learning. Research across a range of subject matter has suggested the efficacy of verbal input combined with nonverbal—typically visual—input. Researchers in second-language acquisition and literacy education (among others) have tested Paivio's dual-coding theory (e.g., Clark and Paivio 1987, 1991; Paivio 1991). In brief, dual-coding theory says that learners are better able to recall information when presented with that information in both verbal and visual modes. The learner's verbal cognition system processes and stores language and the visual cognition system processes and stores images. With both systems simultaneously at work, the learner better retains the information. Teachers' combination of nonverbal demonstrating strategies with verbal describing and explaining strategies took advantage of students' full cognitive-processing capacity.

In their demonstrations, teachers also contrasted correct practice against incorrect practice, better allowing students to understand differences in body position and technique. Teresa used negative demonstration to show Sophia the common errors that she had seen past students make (see excerpt 3.12). Sometimes, after observing students weld or perform some other task, teachers used negative demonstrations to replay students' errors, as Tonya did in her interaction with Simon (see excerpt 3.13). Using negative demonstrations in this way, teachers intervened with instruction that helped students avoid errors.

Similar to teachers' demonstrations, teachers' gesticulations combined with verbal strategies to take advantage of students' capacity to process visual and verbal input simultaneously. With deictic gesticulations, particularly concrete ones, teachers pointed to relevant locations or items in the shared environment, such as when Tonya pointed to the top of Steve's vertical-down weld (see figure 3.7). In addition, welding teachers used iconic gesticulations to illustrate objects and actions, such as when Teresa used an iconic gesticulation to illustrate for Seth how to push the electrode into the joint (see figure 3.9). This chapter discussed teachers' use of metaphorical gesticulations to help convey abstract concepts, such as Teresa's gesticulation for an "umbrella of heat." Metaphorical gesticulations were particularly intriguing because they could visually convey a teacher's verbal metaphor, as when Teresa described "an umbrella of heat" (see figure 3.12). They could also, however, simply illustrate an abstract concept, such as strain.

The metaphors that teachers used in their describing strategies were, on occasion, novel—metaphors created in the moment to convey some phenomenon in the shared environment. As Cameron and Deignan (2003) pointed out, tuning devices tended to precede these novel metaphors. With these tuning devices, teachers seemed to acknowledge the unexpectedness of their descriptions, such as Tonya's unexpected mapping of the beer-belly source domain onto the target domain of a drooping weld (see excerpt 3.6). To interpret such novel metaphors, students first had to be able to match the source to the target domain. But because they were unexpected, teachers' novel metaphors might have created "better-remembered understandings of a given topic" (Grady 2017, 450); that is, novel metaphors may help students recall information.

Much welding jargon arises from conventional metaphors, such as the metaphors of the weld toe and root. In these cases, welders conceptualize the target domain through the metaphorical language, and so to welders, the language seems completely natural—"everyday" in Lakoff's terms—and thus goes unnoticed; toe, root, face, and throat are simply what welders call different parts of the weld. Numerous other examples of metaphorical welding jargon exist, some more novel than others: aging, alligator cut, backhand, blowhole, creep, quenching, stringer, and so on. To help students understand and recall such metaphorical jargon terms, teachers might make explicit their source-to-target mappings. As they learn the specialized language of welding, including its metaphoric jargon, they show themselves to be members of the community.

In sum, this chapter has examined how teachers intervened in students' learning of equipment setup, body position, and technique. Teachers used instruction strategies, particularly demonstrating combined with describing and explaining. In the next chapter, I examine how teachers used less directive tutoring strategies, cognitive scaffolding strategies, to help students develop expert perception—the ability to see, hear, and touch/feel—like an expert.

Chapter 4

Teaching Expert Perception

When I finally accomplished a few solid FCAW[1] welds in the overhead position, I heard my teacher yell from his office, "That sounds real nice, Jo." He could hear me welding because that semester I had chosen the booth that was nearest to his office, about 10 feet away. My teacher didn't even have to look at my welds to see that my overhead skills were (very slowly) improving. He just had to hear me weld. On the flip side was the night that I practiced pulse-spray transfer for the first time. I knew from a Hobart training video that pulse-spray transfer would be louder than the short-circuit transfer that we had practiced for the first half of the semester, and I was definitely generating a loud crackling sound as I welded lap joints and T-joints. So, even though I was generating a lot of spatter, droplets of molten material that fall around the weld, I thought I was perhaps doing ok—that my welds were as strong and clean as they could be. I learned better when a far more experienced classmate, Shane, came over from the adjacent booth to see how I was doing or, rather, as I would learn, to help fix my problem. I told him I thought I was doing ok but admitted I wasn't sure because I had never done pulse-spray transfer before. His response to my optimistic assessment: "Um, yeah, it sounds like shit." Shane, who was further along in the program and who already worked as a welder, knew what pulse-spray transfer should sound like. I, on the other hand, quite clearly did not. This chapter examines the ways in which welding teachers scaffolded students' development of expert perception—senses honed through study, experience, and practice. My FCAW teacher's praise of my overhead welds and Shane's (well-meaning) criticism of my pulse-spray transfer welds revealed their expert hearing and, too, my own lack of it.

Although welding teachers used a range of tutoring strategies as they scaffolded students' progression toward expert perception, I show here that, in particular, they employed cognitive scaffolding strategies for this purpose (see table 4.1). With cognitive scaffolding strategies, teachers invited intersubjectivity. And, especially with pumping questions, welding teachers diagnosed students' level of understanding and thus helped teachers make their responses contingent and within students' ZPD. Teachers' open-ended pumping questions, in particular, facilitated interactivity in that they pushed students to contribute to the interaction.

Teachers in the skilled trades have developed expert perception, a component of embodied knowledge; they see, hear, and touch/feel as experts in welding.[2] One of their main goals, beyond teaching equipment setup, body position, and welding technique, is to move students forward on the path toward expert perception, making it possible for students to assess a welding environment as a member of the welding community of practice.

As Goodwin (1994) and Ingold (2001) have pointed out, the first step in helping a student acquire expert perception is making salient a specific phenomenon in the shared environment, making it a figure against ground (Goodwin 1994, 606). Goodwin called this step in teaching "highlighting." In discussing the need to separate a phenomenon from its environment, Goodwin focused on learners' sense of sight. While learning to see as an expert is critical, in welding other senses must develop too. Most notably, welding students grow in their ability to differentiate the sound of a good weld from a bad weld in a given process and position. By highlighting phenomena, teachers build intersubjectivity and thus help ensure a shared understanding of the task at hand.

Besides identifying a phenomenon in its environment, a teacher must also help a student transform the phenomenon into what Goodwin (1994)

Table 4.1. The cognitive scaffolding strategies found in this study's data

Strategy	**Definition**
Pumping question	Teacher asks a question that gets a student to respond. Pumping questions vary in the extent to which they constrain a student's response; they can be open-ended or closed.
Referring to a previous topic	Teacher refers back to the earlier topic or occurrence of an issue.

called a "knowledge object" (606). The student must learn to interpret the phenomenon; they must learn what the phenomenon signals, what it means. Goodwin discussed this critical step in development as "coding" and, focusing on the sense of sight, gave the example of a Munsell chart for classifying dirt by color. Using this example, identifying a given color constitutes highlighting, but understanding what that color means (e.g., the presence of iron oxides) is coding. In relation to welding, identifying a recess at the side of a weld bead is highlighting; interpreting that recess as undercut is coding.

Seeing as an Expert

Prior research has examined the teaching and learning of the expert sight in a variety of workplace settings, particularly in medical settings where its importance is readily apparent. Studying the impact of routine on an interaction during a cholecystectomy, the removal of the gallbladder, Koschmann et al. (2011) examined how a surgeon in training, a faculty member, and a third-year medical student came to a shared understanding of an object, in this case, the cystic artery, in the removal of a gallbladder. Their study exemplified how old timers help newcomers identify and code objects and thus become more expert members of the community of practice. Similarly, Mentis, Chellali, and Schwaitzberg (2014) studied how surgeons teach the "indirect vision" of surgery during dissections and thus help trainees develop "an effective medical gaze" (3). Like the surgeon trainees in these studies, welding students must learn to identify a relevant phenomenon in the visual field, for example, an overlapping weld, and to code its potential meaning, for example, a too-slow travel speed.

Because the ability to visually inspect welds is critical, welding teachers prioritize development of students' visual perception. Resources for learning how to weld, including the Hobart manuals and videos used at DMACC and in other programs, cover visual inspection of welds for defects and discontinuities. A discontinuity interrupts the structure of the weld but falls within acceptable margins of error. Discontinuities are less serious than defects. A defect is a discontinuity that has developed into a flaw that puts a weld at risk of failure. Therefore, a defect makes a weld unacceptable and necessitates a repair (Swanton Welding Company 2016; The Welding Master 2017). Defects can be internal or external. For example, incomplete weld penetration is an internal defect. Undercut is an external defect. While visual inspection of welds for discontinuities and defects constituted part of students' periodic skills assessment, visual inspection was ubiquitous through every class period as teachers came to students' booths to watch them weld or to inspect work they had just completed, and as students brought their work to teachers for evaluation.

For example, Tom visited Stephanie's booth on his tour around the lab to check on students' progress. That day, Stephanie was working on spray-transfer GMAW for the first time (like I was the day that Shane told me my welds sounded "like shit"). To diagnose her understanding, Tom observed as Stephanie welded. After watching this first attempt, Tom decided that demonstrating the correct technique would be helpful, and he switched places with her. As he started his demonstration, he noticed a problem: big drops of filler metal were coming from the gun, not a fine spray. Tom used the opportunity to help Stephanie learn how to identify the problem, to code its meaning, and then to adjust the machine settings. To push forward her ability to perceive the problem, Tom used pumping questions.

Welding teachers' pumping questions differed in the degree to which they constrained students' responses. Some pumping questions, particularly yes-no questions, required just minimal responses from students. Oftentimes, welding teachers' yes-no pumping questions asked students to confirm that they had noted some phenomenon in the environment, such as too-big drops from the gun. Open-ended pumping, such as "What do you think?" allowed for a broad range of potential responses and facilitated more substantive responses as well.

Tom highlighted the problem of the big drops with a yes-no pumping question. With this question, he generated some intersubjectivity; he and Stephanie were attuned to the same phenomenon. In addition, with his pump question Tom diagnosed Stephanie's understanding; she had perceived what he had highlighted. The pumping question also pushed Stephanie to respond, albeit minimally, and thus increased the interactivity of the exchange:

Excerpt 4.1

TOM:	Now, we want to be 90 for a work angle and then a 10-degree push. Depending on the angle we're working on. [4s] Ok. So do you see those big drops heating off?
STEPHANIE:	Uhhuh.

Assured that Stephanie also saw the big drops, Tom stopped to address the problem. First, he coded the meaning of the big drops for her. Their settings were not high enough for spray:

Excerpt 4.1 (continued)
TOM: We're not at spray yet. I'm going to stop. [2s]

Tom's next move, after highlighting and coding, was to offer a solution—adjusted machine settings. Tom directed Stephanie to turn up the voltage and the wire speed, using a concrete deictic gesticulation to draw Stephanie's attention to the machine's control panel (figure 4.1):

Excerpt 4.1 (continued)
TOM: Let's- let's go 28 on voltage [2s] and we'll take the wire speed all the way up to 400. [5s] Ok. Ready?[3]
STEPHANIE: Yep.

Tom began demonstrating again, using describing strategies to highlight the characteristics of the spray that he saw and also his speed as he welded:

Figure 4.1. Tom used a concrete deictic gesticulation when telling Stephanie to adjust the voltage and wire speed on the welding machine.

Excerpt 4.1 (continued)

TOM: [4s] That's more of what we want. You don't see those big drops. We're getting them every once in a while. [3s] But also notice how slow I'm going. [7s]

After Tom stopped his demonstration and raised the face-shield of his helmet, Stephanie asked a question to confirm what Tom had said about travel speed across the weld. Her question verified that she had isolated the figure of his travel speed from the ground of the rest of the shared environment. As a contingent response to Stephanie's understanding, Tom described the results further:

Excerpt 4.1 (continued)

STEPHANIE: So it's not fast?

TOM: Nope. Like but you're getting really big welds. [So it

STEPHANIE: [Ok.

TOM: doesn't look like you're moving quickly, but you're filling this weld up very very quickly. So it will actually take much less time than a short-circuit weld.

In his response to her question, Tom used another cognitive scaffolding strategy: referring to a previous topic. In this case, Tom compared the appearance of pulse-spray transfer welding to short-circuit welding, the process that Stephanie and her fellow GMAW classmates had just learned. With this strategy, welding teachers associated the task at hand with a task that students had already been exposed to, if not mastered. In other words, teachers related new information to old information, scaffolding students' learning step-by-step by making it contingent upon what they already knew. Referring to a previous topic could also help teachers generate intersubjectivity in that it identified a shared common ground for understanding the new task.

But Tom also determined at that point that the settings needed further adjustment. Before making those adjustments, Tom described what he had seen and had not seen coming from the gun, highlighting for Stephanie the characteristics of the results they were getting and describing the results they should be getting. After, Tom once again coded their outcomes, using an explaining strategy to convey what this meant: The settings were still too "cold," and they needed to turn the voltage up yet again. Tom described his actions once again ("I'm going to shift it a little bit"), welded again to see if he got better spray, and then coded the result ("Well that's too much"):

Excerpt 4.1 (continued)

TOM: Let me just double check this. [Bends down to adjust the settings on the welding machine] Because it's- I still feel like we're not quite right. [7s] It all looks good. I'm trying to figure out- We're not quite getting that full spray. [Looks again at his demonstration weld] I'm still seeing those bigger drops come out from time to time. [Turns back to the welding machine, 3s] Cold. [Adjusts the settings on the welding machine again, 14s] Let's see if that works. [Turns back to the weld table, 5s] I'm going to shift it a little bit to help me see. [Starts welding again, 5s] Well that's too much.

After finding that turning the voltage up from 28 to 30 was too much, Tom adjusted the voltage setting yet again. But before welding again, Tom paused to check to see whether Stephanie had been following his descriptions of his actions and his explanations of what those actions meant. He stopped, raised his face shield, and asked Stephanie another pumping question aimed at diagnosing her understanding of his machine-setting adjustments:

Excerpt 4.1 (continued)

TOM: [Raises his face-shield, 4s] So you can see when I crank that up to 30- You see how much wider that arc got?

STEPHANIE: Right.

TOM: That's because the arc got longer.

STEPHANIE: Ok.

As he spoke his pumping question, Tom moved his hand in a short back-and-forth movement across the top of the weld, a metaphoric gesticulation that captured the concept of arc width (figures 4.2 and 4.3). His pumping question combined with his gesticulation clarified what had just occurred, the result he had obtained by adjusting the settings. After Stephanie confirmed that she had indeed seen what Tom had highlighted, that the arc had become wider ("Right"), Tom offered yet another explanation, or coding, of the visual field ("That's because the arc got longer").

Then, Tom moved to application, describing and explaining what he planned to do to the settings and why he planned to do it:

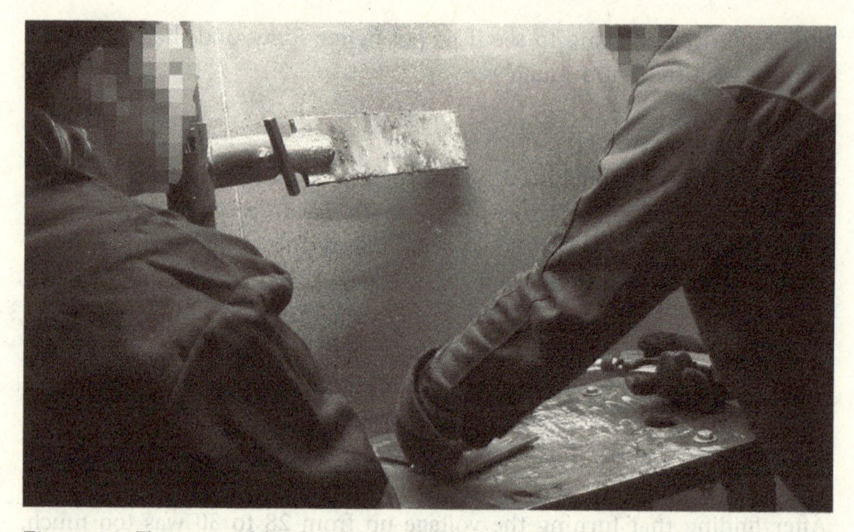

Figure 4.2. Tom moved his hand back and to the left at the beginning of a small back-and-forth movement across the weld to indicate the concept of arc width.

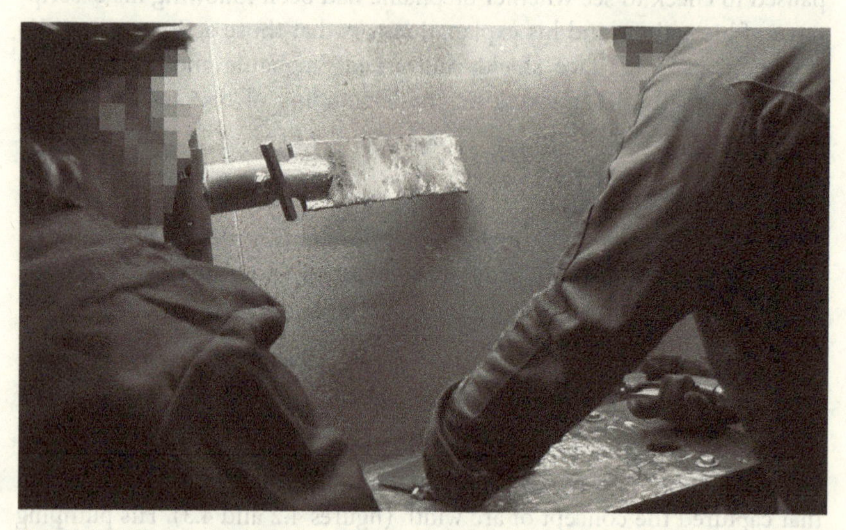

Figure 4.3. Tom moved his hand forward and to the right to finish a small back-and-forth movement across the weld to convey the concept of arc width.

Excerpt 4.1 (continued)

TOM: That's with the voltage control. [Prepares to weld again, 2s] So I'm going to try and shrink that down so we have a little more control in the groove. [Starts welding again, 3s]

Finally, Tom had sized the arc as he wanted it. He highlighted for Stephanie that the mission was accomplished ("There it is") before describing once again his slow speed across the weld, even asserting as fact that Stephanie could see his actions:

Excerpt 4.1 (continued)

TOM: There it is. [2s] So you can- You can see how slow I'm moving. [3s] We want it nice and smooth. [2s] The goal is that the top of our weld- the face is flat. [5s]

In wrapping up what had turned into a lesson in expertly seeing the arc and spray and determining changes to the settings based on that seeing, Tom described the appearance of the ideal spray-transfer weld: smooth and flat. Throughout this lesson in machine setup, Tom's pumping questions kept Stephanie at least minimally engaged by pushing her to confirm that she could see each phenomenon that Tom wanted to highlight. With them, Tom made his demonstration contingent upon Stephanie's perception of the visual field.

Scaffolded teaching of expert seeing occurred also as teachers evaluated students' welds. In such interactions, teachers "read" the welds that students had just made, using students' work to diagnose problems and to direct them in machine settings, body position and, often, technique. After demonstrating a vertical-down lap joint earlier that day, Tonya returned to Steve's booth to check on his progress after he had tried the same joint by himself. Tonya began with a pumping question aimed at diagnosis, specifically, ensuring that Steve understood that for this vertical-down weld, he should angle the gun upward at a 35- to 45-degree angle ("Are you pointing up?"). When Steve answered affirmatively, Tonya concluded that Steve's results thus stemmed from another problem: Steve's travel speed had been too slow. She used a concrete deictic gesticulation to highlight the location on the weld that bulged (figure 4.4) and another pumping question to code its meaning:

Excerpt 4.2

TONYA: Are you pointing up?

STEVE: Yeah.

TONYA: You're just traveling too slow. See how the puddle's getting ahead of you?

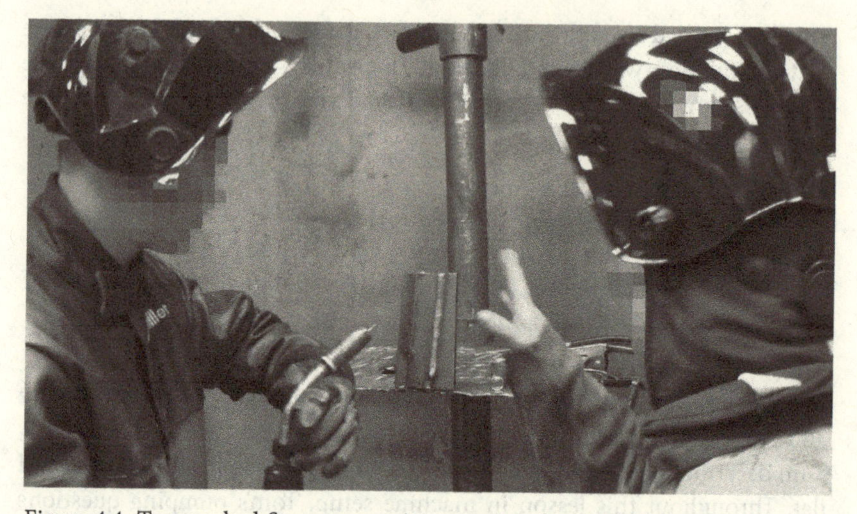

Figure 4.4. Tonya asked Steve a pumping question that pushed Steve to interpret the meaning of the physical characteristics of the weld that he had just produced: "See how the puddle's getting ahead of you?"

In asking for coding rather than highlighting, Tonya's pumping question differed from the ones Tom had used as he adjusted the voltage for Stephanie (excerpt 4.1). Tom's pumping questions highlighted physical characteristics in the environment and asked Stephanie to acknowledge that she too could see them. Then, Tom followed up with coding—explanations of the phenomena. Tonya's pumping question instead asked Steve to read, that is, to code the weld. It asked him not just to perceive the bulge in the weld but to interpret its cause.

Although teachers' pumping questions often requested simply that the student acknowledge perceiving the phenomenon that the teacher highlighted, they could generate more substantive responses from students as well and thus greater interactivity. Such an interaction occurred when Teresa visited Saul's booth again after having fixed a problem with the gas flowing to his booth. This time, Teresa turned the voltage on Saul's machine down and then observed as he performed a vertical-up SMAW weld. After Saul finished, the two looked at his weld together, reading it to determine whether the adjustment in voltage had improved Saul's results. Teresa began by highlighting observable phenomena in the weld, and Saul, for his part, extended her description:

Excerpt 4.3

TERESA: All right for starters it didn't look like your keyhole was getting too big on you.
SAUL: No I- I think maybe- maybe a little-
TERESA: Right towards the end a little bit.
SAUL: A little cold if anything.

As mentioned in chapter 3, Saul was nearing graduation, and he already worked as a welder. That he had already developed some expert vision showed in his understanding and extension of Teresa's description of the weld's keyhole. As Saul looked at the backside of the weld (figure 4.5), Teresa began a series of pumping questions that called upon Saul to code the physical characteristics of the weld he had just produced. Her first pumping question asked Saul to assess the quality of the weld's penetration through the root opening:

Excerpt 4.3 (continued)

TERESA: Yeah. How was the penetration?
SAUL: Not that great. [Laughs]
TERESA: Ok. [Bends the welded coupons back, 2s]
SAUL: We got it for a little bit.

Figure 4.5. Saul looked at the backside of the weld to assess the weld penetration.

At the moment of her question, Teresa had not yet seen the backside of the weld to know for certain the quality of the weld's penetration. She gave Saul an opportunity to evaluate his results on his own, a time to see the weld with expert vision. Her open-ended pumping question made room for a more substantial response. In answering her, Saul laughed as he replied, "Not that great."

Even after Teresa bent the weldment back so that they could examine the penetration of the weld together, Saul extended his assessment. Not beating himself up for a weld lacking continuous penetration, he noted, "We got it for a little bit," and used a concrete deictic gesticulation to indicate the section of the weld that showed better penetration (figure 4.6). His extended assessment of his own work reflected his experience; he was able to highlight a phenomenon (a section of the weld) and code its meaning.

Teresa quickly agreed ("Yep"), and then used a yes-no pumping question, paired with a concrete deictic gesticulation (figure 4.7), to highlight the unevenness of the coupons:

Excerpt 4.3 (continued)
TERESA: Yep. See how uneven these plates are?
SAUL: Yeah.

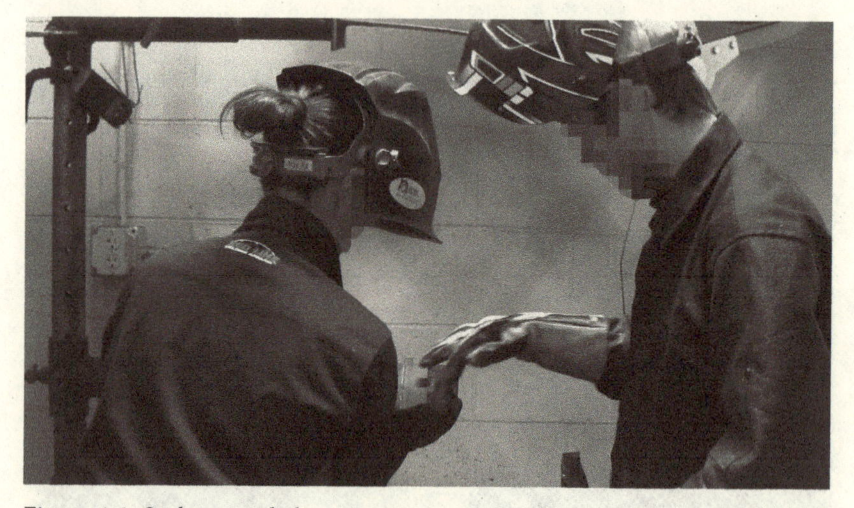

Figure 4.6. Saul assessed the penetration on the weld he had just performed, noting, "We got it for a little bit." He used a concrete deictic gesticulation to highlight the section of the weld with better penetration.

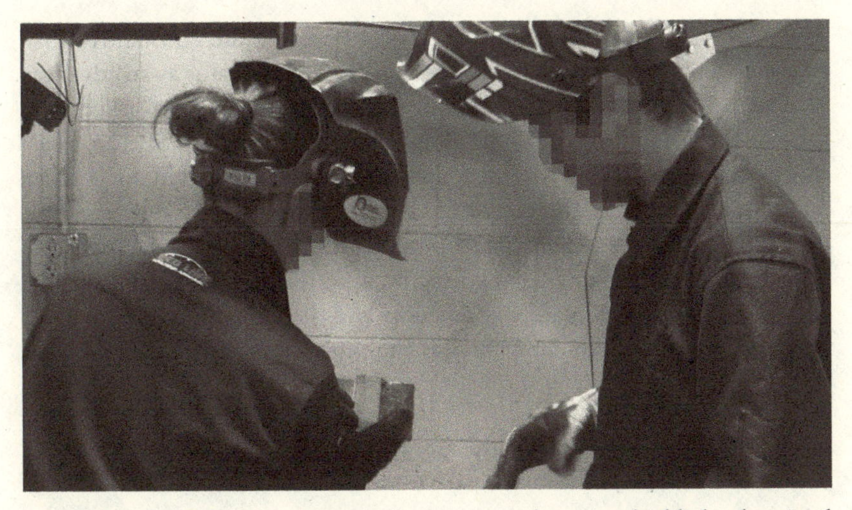

Figure 4.7. Teresa used a concrete deictic gesticulation to highlight the visual phenomenon she wanted Saul to notice—the unevenness of the coupons.

The unevenness of the coupons tacked together likely played a role in the less-than-complete penetration that Saul achieved. Considering these results, Teresa asked yet another pumping question, this one confirming his earlier coding of his results—his weld had been "cold":

EXCERPT 4.3 (continued)
TERESA: That probably didn't help much but, you know right up to here- It was a little cold huh?
SAUL: Yeah and I- I tend to like to run right on that edge of-

Beginning what seemed to be an assertion that he typically likes to run at the top of the typical voltage range ("I tend to like to run right on that edge of-"), Saul received an acknowledgment from Teresa that the voltage she had chosen was a bit too low, and she directed him to turn it up just slightly:

Excerpt 4.3 (continued)
TERESA: All right so let's turn it up. I turned it down to 17. I want you to try it at 17.2.
SAUL: Two?
TERESA: Yep. [2s] Um you're really close to having it.

Saul's contributions throughout excerpt 4.3 were more substantive than those of Stephanie or Steve, likely due in part to Saul's status as an advanced student but also due to the pumping questions that Teresa asked him. One of those pumping questions was open-ended, asking Saul to assess the weld's penetration. Another, a yes-no question, sought to confirm Saul's own coding of the weld's physical characteristics ("It was a little cold huh?). However, even closed pumping questions, such as Tom's and Tonya's, engendered intersubjectivity by asking students to confirm that they too saw a particular phenomenon and also generated interactivity by pushing students to contribute to the interaction.

Hearing as an Expert

Although welders certainly depend upon their sight as they weld and as they evaluate the quality of a weld, they also listen while welding to determine whether the machine settings and technique are correct. Indeed, in online forums, professional and amateur welders' comments reflect the importance of auditory feedback. For example, on the Hobart discussion forum, Dan (2004) responded to sound clips of GMAW that another user had posted:

> The sound could just be distorted, but that sounds more like the crackle sound, which means the arc length is a little short. They need to either increase the voltage a little or reduce the wire speed slightly. When you hear the crackle sound you tend to end up with a slight amount of spatter. I like to take the arc to the crackle sound then just slightly lengthen it until the crackle is gone, and you hear the hissing sound instead.

In forums that companies such as Lincoln, ESAB, Miller, and Hobart maintain, as well as the Welding Design & Fabrication forum, the Welding Tips and Tricks forum, and others, learners frequently describe the sounds of their welds and ask for interpretation, or coding, from more-expert users. For example, in response to fishbum's request for help with GMAW welding of aluminum, Gunney (2013) responded:

> Welding by the sound is a rather subjective measure, but once you get a machine dialed in for a good weld on aluminum,

> the sound is a good indicator to let you know how the weld is doing. In other words, get a good, clean weld with good penetration and get used to the sound that is produced, then the sound will help guide you in changing settings for different material thickness, etc. Having said that, I get more of a crackle than a hiss when I weld aluminum, but the crackle is a very even crackle. If the crackle is intermittent or has a lot of "pops" then I'm usually seeing more spatter. I do know people who weld aluminum with a standard mig process that get more of a hiss, but they have their voltage and wire feed turned up real high and their resulting travel speed is crazy fast.

Gunney's response, like Dan's (2004), used some of the onomatopoetic terms—"hiss," "pop," and "crackle"—that welders use to describe the auditory input that they perceive while welding. Gunney's post, in particular, shows how welders attempt to associate sounds with outcomes, machine settings, and techniques. Welding teachers, including my own, frequently referred to the sound that welds make, using words like "buzzing" and "humming." My GMAW teacher, for example, pointed out that a good weld sounds like bacon sizzling in a pan. Many other welders have used this analogy (e.g., Eastwood n.d.).

In other fields in which the ability to discern via hearing is critical, researchers have examined the development of auditory acuity. For example, research in music education has tested various methods of developing the ability to identify errors in intonation and pitch (Byo and Sheldon 2000; Gonzo 1971; Scherber 2014; Sheldon 2004), melody (Beckman 2014; Talbert 2012; Thornton 2008), and rhythm (Byo and Sheldon 2000; Sheldon 2004). Research on L2 learning has investigated the acuity of automatic-error-detection software (Qian, Meng, and Soong 2012; Strik et al. 2009). L2 researchers have examined methods of developing students' listening skills (Goh and Hu 2014; Siegel 2013; Vandergrift and Goh 2012; Vandergrift and Tafaghodtari 2010). In general, such research has shown that with scaffolded teaching from a more-expert other, students can develop their auditory acuity.

Research within the field of welding, though rare, has taken up welders' ability to perceive changes in arc sound (e.g., Arata et al. 1979).[4] Tarn and Huissoon (2005) specifically tested welders' reliance on sound through a variety of psychoacoustic experiments. They found that welders needed the auditory input of welding in order to produce good welds.

However, they also found that welders were eventually able to rely on visual input to compensate for a lack of auditory feedback. They also found that delayed auditory feedback at 400 milliseconds was "enough to cause completely unstable control conditions in human welders" (1116). This study supported welders' long-held view that the sound of a weld is an important indicator of weld quality.

A prototypical example of interpreting sound occurred when Tonya began a demonstration of oxyacetylene welding for Stewart. Oxyacetylene welding differs from electric processes in that it relies solely on a mixture of oxygen and acetylene gases to fuel the welding torch. It doesn't take long before the tip of the torch gets clogged with soot. Immediately after lighting the torch and adjusting the mixture of oxygen and acetylene to achieve a neutral flame, a flame in which the gas mixture creates a small white cone with no "feather" from excess acetylene (figure 4.8), Tonya used a pumping question to determine whether Stewart perceived the sound of the flame:

Excerpt 4.4

TONYA: Ready?

STEWART: Yep.

TONYA: The biggest problem I have with this one, is remembering to keep the 45-degree angle into the joint.

STEWART: Ok.

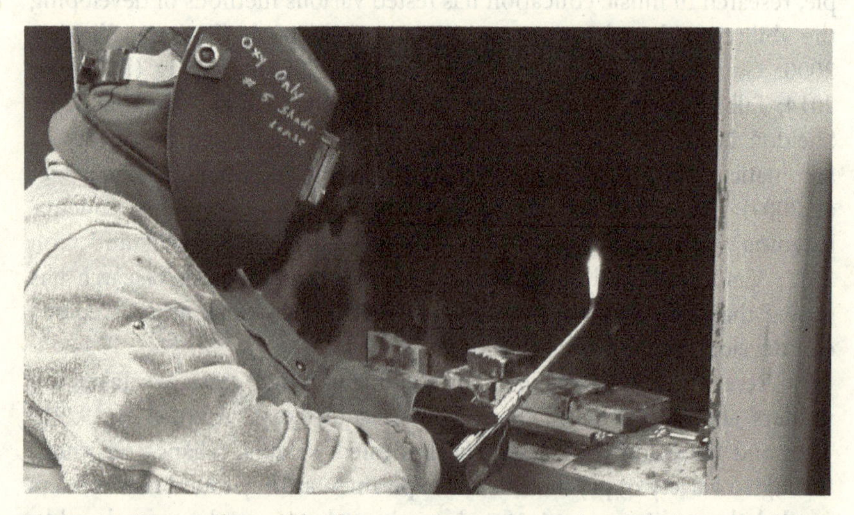

Figure 4.8. Tonya adjusted the flame on the torch and then asked Stewart, "Hear how loud that is?"

TONYA:	If you don't do that- [Lights torch] Can you turn the light on for me? [Points to light; Tonya adjusts the gas mixture, 12s] Hear how loud that is?
STEWART:	Yeah.
TONYA:	[Turns off torch] Tip needs cleaning.

A dirty tip on an oxyacetylene torch can show itself in other ways. It usually generates a misshapen flame, as I learned (over and over) in my oxy welding class. But as Tonya observed, it can also generate a louder sound of gas through the torch tip. Tonya's pumping question highlighted the sound and generated some interactivity from Stewart. And after, Tonya coded the auditory phenomenon: The torch tip had clogged up.

Similarly, in his demonstration of spray-transfer GMAW, Tom moved from a pumping question that highlighted an auditory phenomenon, popping, to an explanation strategy that coded it for Stephanie:

Excerpt 4.5

TOM:	Now you heard that popping?
STEPHANIE:	Yeah. I was going to ask what that is.
TOM:	Yep, so that's generally a little bit of impurity entering where the arc is. In a weld like this it is most likely, coming from the underside of the bevel. [2s] So it's probably we're melting away enough of that edge that we're starting to pick up just a little bit of mill scale- Gets pulled into the weld and causes that popping. Ok?

The pumping question drew Stephanie into the interaction, highlighted the sound, and made way for Tom's fairly detailed explanation, his coding, of the sound.

A final example comes from Teresa's demonstration of welding an open-groove joint. Like Tonya and Tom, Teresa used a pumping question to confirm that her student perceived an important auditory phenomenon (figure 4.9). In this case, however, the phenomenon was wanted; the whooshing sound occurs when the keyhole has opened and weld penetration can occur:

Excerpt 4.6

TERESA:	So. Angle, straight 90. You have to be right in between the plates the whole time. [Flips hood down and begins welding]
SOPHIA:	Ok.

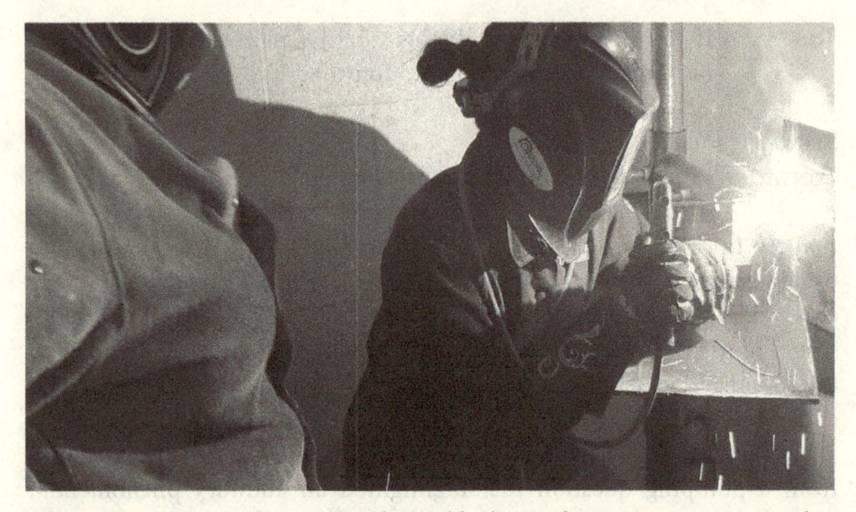

Figure 4.9. As Teresa demonstrated a weld, she used a pumping question that highlighted an important auditory phenomenon: "Can you hear that whooshing sound?"

TERESA: Can you hear that whooshing sound?
SOPHIA: Yeah.
TERESA: That's telling me that I'm getting through to the other side.
TERESA: See how I'm only moving a little bit forward?
SOPHIA: Yeah.
TERESA: Basically just digging out between the plates so I can continue my weld. I'm coming back and I'm pushing through.

Together, these three cases reveal how welding teachers used pumping questions to ensure that students perceived the sound that teachers highlighted, creating intersubjectivity, and subsequently interpreted the meaning of those sounds, including actions to take, such as cleaning the torch tip.

Feeling as an Expert

Throughout my welding education, gas canisters gave me some anxiety. You turn the knob on top of the gas canister to let the gas flow, and then you adjust the pressure on the regulator. But it's not that simple. You open combustible gases, such as acetylene and argon-CO_2 blends, just a quarter turn or so, but you can open the oxygen, a noncombustible gas, all the way. But oxygen canisters open to the right, unlike combustible gases. In my GMAW class, my teacher demonstrated for me what a quarter turn looks like, but then he closed the canister and told me to try it myself. The idea was that I would feel what a quarter turn of the knob feels like. I needed the same sort of scaffolding when using a 1⅛ wrench to install a regulator on a gas canister. I needed to tighten the nut that held the regulator on the canister, but if I turned the wrench too much I could strip the fitting's threads. In this section, I discuss how welding teachers—like my GMAW teacher—scaffolded students' perception of touch or feel.

Research in medical training has investigated the role of expert touch in embodied knowledge. Hindmarsh and Pilnick (2007), for example, studied the development of what they called "intracorporal knowing" during a preoperative intubation procedure involving an anesthetist and an operating-department assistant. They found that slight differences in pressure on a colleague's hand or differences in presenting an instrument helped coordinate the participants' actions. In later research, Hindmarsh, Reynolds, and Dunne (2011) examined what Becvar, Hollan, and Hutchins (2008) called "professional touch." The Hindmarsh, Reynolds, and Dunne (2011) study focused on student dental clinics (i.e., dentistry training), where demonstrators supervised students (dentists in training) who were "learning to appreciate the tactile nature of dental phenomena" (491). Such studies point to the critical importance of tactile perception in the learning of embodied knowledge. In relation to welding education, they are applicable to situations like learning how to put a regulator on a gas canister.

In addition to research on expert touch, ergonomics research is also relevant to the teaching of the embodied knowledge of welding. Part of learning how to perceive like an expert welder is being able to recognize a comfortable body position. Comfort is an important consideration given that welders typically work eight or more hours per day, sometimes in physically challenging spaces, and must therefore conserve energy and minimize discomfort. In a study of work-related musculoskeletal disorders, Singh and Singhal (2016) reported that common body positions such as kneeling and welding positions such as overhead can generate a variety of muscle disorders (265). More specifically, Rahman et al. (2015) showed that welders suffer from a variety of work-related problems, such as back injury, shoulder pain, tendonitis, knee pain, and carpel tunnel syndrome. Welding blogs and forums stress the importance of being comfortable no matter the welding position or process. Responding to a question from a novice welder, cj737 (2018) advised about GTAW, "To weld well, in any process, you must be relaxed. You can be braced and still relaxed . . . or even standing on one foot with the other hovering over a TIG pedal and remain relaxed. It's how you will find your natural physical balance. To avoid fatigue, you want to prop somehow." About SMAW, Paladin (2011), wrote, "I agree learning to position your body, heads, and hands, so you can see, be balance[d] and comfortable, and hands free to manipulate the rod is important in learning to weld." Writing to someone considering a career in welding, Jeffrey Grady wrote, "Then there is the comfort level. How physically comfortable do you feel you need to be. I say this because even in school there are certain welding positions that require you to endure a certain level of discomfort" (Grady 2008). In short, welding can involve crouching and stretching and other body positions that are difficult to maintain for a long time. Welders need to learn to get comfortable, particularly when performing out-of-position welds. When I learned overhead FCAW, my teacher told me time and again to rest my arm against the wall or against the arm of the weld table to steady myself and to conserve energy for repeated overhead welds.

Pumping questions, along with demonstration strategies, played a strong role in how teachers scaffolded students' learning of body position. After observing Spencer as he welded in the overhead position, Ted diagnosed a problem with the student's body position. To intervene, Ted demonstrated to Spencer what he had been doing wrong. Like Tonya in excerpt 3.13, Ted used demonstration to replay a student's actions. Ted held his arms out straight from his body, exaggerating the position that he had seen Spencer take during his overheld weld (figure 4.10). In addition to replaying Spencer's position, Ted told Spencer to take the same position, highlighting for Spencer the inefficiency of the position. When Spencer had raised his arms, Ted employed a pumping question that asked Spencer to consider what it would take to hold his arms in that position over the course of a day's work:

Excerpt 4.7

TED: I'm going to give you the life vest one oh one. So hold your arms out like this. Ok. Now. Hold your arms out as long- How long you going to hold them like that?

SPENCER: Not for very long.

TED: Ok.

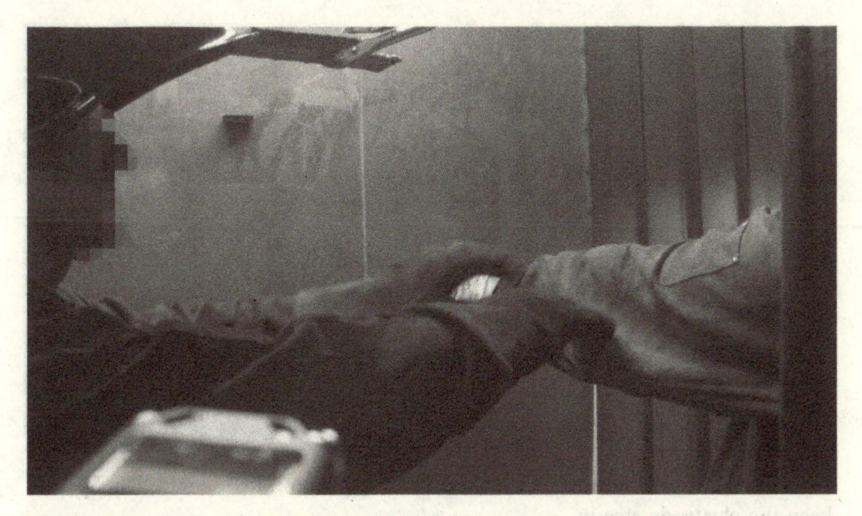

Figure 4.10. Ted told Spencer to extend his arms in an inefficient body position—one that Spencer had used just before when Ted watched him weld. Ted then asked a pumping question to get Spencer to think about that body position.

Ted's pumping question pushed Spencer to articulate what he felt through his outstretched arms—that it would be very difficult to sustain that body position. And, like other pumping questions, it had the added benefit of bringing the student into the interaction as an active participant, if only briefly.

A minute later, Ted returned to this topic of Spencer's outstretched arms to point out that the body position, besides being unsustainable, had made Spencer's arms unsteady and shaky. Again using demonstration, Ted contrasted an ideal body position to Spencer's. First, he showed Spencer the ideal, using telling strategies to highlight salient features of the steady body position (figure 4.11) and using a pumping question that pushed Spencer to concede that the electrode did not "jump around" when held in Ted's tucked position:

Excerpt 4.7 (continued)

TED: But look where my body is. Look where my leg and my hip is.

SPENCER: You're pushed up against the table so-

TED: Yeah, everything's, snug. See that rod jumping around?

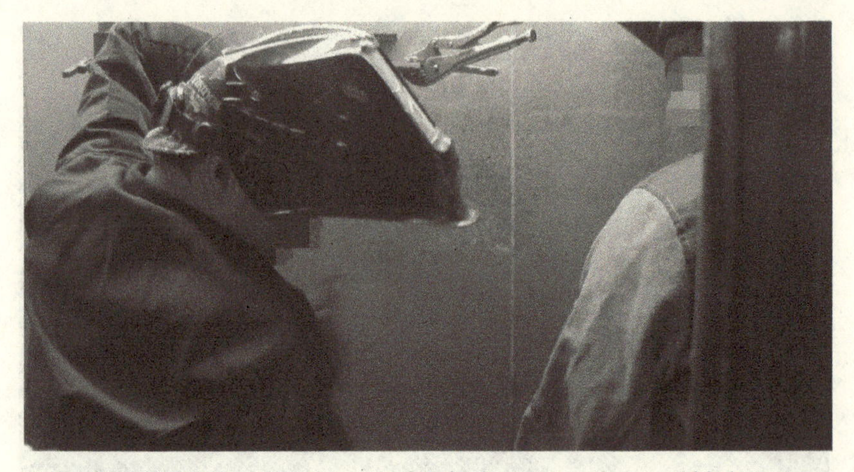

Figure 4.11. Ted demonstrated a tucked-in body position that would help Spencer keep the electrode steady.

SPENCER:	[Shakes his head]
TED:	No.

Then, Ted used negative demonstration again to replay Spencer's body position. This time, he held an electrode upright (figure 4.12) to contrast the electrode's behavior—swaying back and forth—when held in stretched-out arms:

Excerpt 4.7 (continued)

TED:	You're doing this-
SPENCER:	Not steady.
TED:	Absolutely. I can't- I mean even at the best it's vibrating back and forth.

In response to Ted's demonstrations, which had combined with telling strategies and pumping questions, Spencer coded the meaning, "Not steady."

When teaching Stephanie to weld around a pipe, Tom tried to help Stephanie feel a body position that was comfortable and allowed for consistent welds. Like Ted, Tom wanted to help his student feel the difference between comfort and discomfort. After all, like Spencer, Stephanie did not know what comfortable felt like. She only knew that she was having difficulty achieving a consistent weld. After demonstrating and describing an

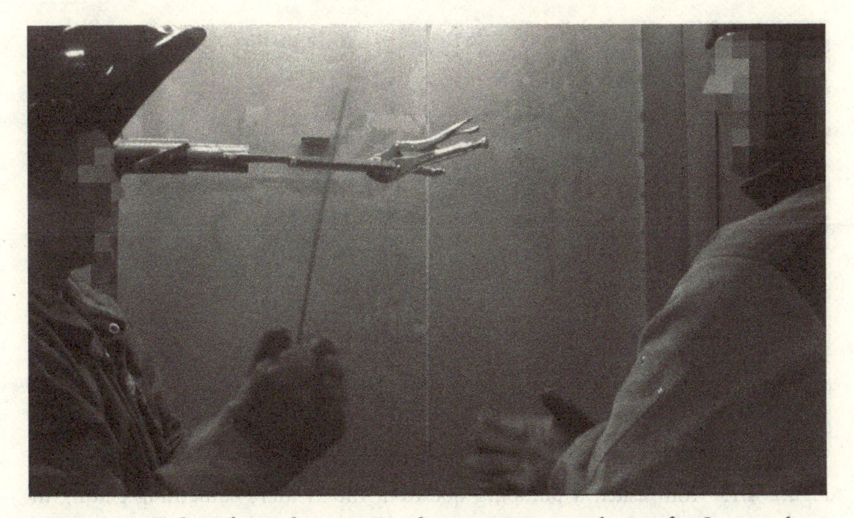

Figure 4.12. Ted performed a negative demonstration, replaying for Spencer how the electrode had swayed back and forth when Spencer had welded with his arms outstretched.

effective movement around the pipe (excerpt 3.4), Tom told Stephanie to do a dry run of the pipe weld so that she could feel her level of comfort or discomfort throughout the weld's stages. As Stephanie moved toward the bottom of the pipe in her rotation around it (figure 4.13), Tom asked his first pumping question. His open-ended question pushed Stephanie to consider the placement of her arms and the welding gun in relation to the pipe:

Excerpt 4.8

TOM:	Do a dry run.
STEPHANIE:	Well, I don't-
TOM:	Ok.
STEPHANIE:	Down.
TOM:	Yep. So is there something getting in the way?

When Stephanie made a face that indicated she did not know how to answer his question, Tom sharpened the focus of his inquiry, asking a second pumping question that scaffolded Stephanie's learning by further constraining her possible answer. Rather than asking whether "something" was "getting in the way," he asked more specifically, "Did you feel tension in your hand?" As he asked the question, he negatively demonstrated the

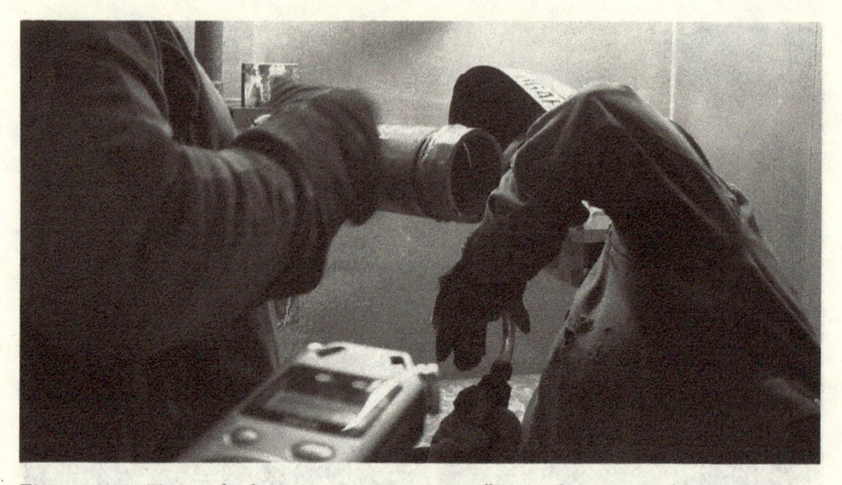

Figure 4.13. Tom asked a pumping question, "So is there something getting in the way?"

sharp bend of the left wrist that would likely generate discomfort, replaying Stephanie's position (figure 4.14):

Excerpt 4.8 (continued)
STEPHANIE: [Makes a face indicating she doesn't understand the question]
TOM: Did it feel- Did you feel tension in your hand?

Given that Tom was making a point of helping Stephanie understand body position and was clearly interested in the angle of her hand and wrist in relation to the gun and the weld surface, Stephanie understood that Tom expected her to answer "yes." But when Tom followed up with a pumping question that pushed her to consider the exact location of the tension in her hand, wrist, and arm, Stephanie could not identify her own discomfort:

Excerpt 4.8 (continued)
STEPHANIE: Yeah.
TOM: Where was the tension?
STEPHANIE: I don't know.

With a concrete deictic gesticulation, Tom identified the exact spot he meant (figure 4.15), simultaneously articulating the location by touching Stephanie on the wrist:

Figure 4.14. Tom asked, "Did it feel- Did you feel tension in your hand?"

Excerpt 4.8 (continued)

TOM: What I saw- I was expect- I would expect it to be in that wrist.

STEPHANIE: This one.

Figure 4.15. After Stephanie failed to answer his pumping question, Tom provided an answer himself: "What I saw- I would expect it to be in that wrist."

After Tom provided an answer, Stephanie generated further interactivity by pointing out which wrist had experienced the tension, indicating with "this one" that she understood the place on her body that she should attend. Tom then told her to do another dry run, this time with her left hand on the top of the gun:

Excerpt 4.8 (continued)
TOM: Flip that hand over on top of the gun. Try again.
[Stephanie tries again, 2s]
STEPHANIE: Oh, that would be a lot easier.

After Stephanie indicated that she could feel the difference between the new, comfortable position and the position she had been using, Tom continued his lesson in identifying body comfort, asking yet another pumping question aimed at getting her to notice her right arm. Again, he used telling strategies immediately after his pumping question to correct her position and enable her to feel a more comfortable position. This time, however, Tom scaffolded Stephanie's learning by physically guiding her movement, placing his hands on the gun and rotating it as Stephanie performed the movement again (figure 4.16):

Excerpt 4.8 (continued)
TOM: Ok? The other thing is- Start at the top and work through it one more time. Stop when you get to the bottom. [3s]
Ok? You notice how you're really having to push this hand out? So go back like you're on the bottom of the pipe. [2s]
If you take this, and rotate it. [2s]
STEPHANIE: Oh.

After they completed the motion together, Tom asked a yes-no pumping question, this time explicitly asking Stephanie about comfort:

Excerpt 4.8 (continued)
TOM: Isn't that more comfortable?
STEPHANIE: So go like, that?
TOM: Yep. That way this arm isn't pushing out. You can try and keep it in tight, and it's just- You sort of roll as you come down.
STEPHANIE: Ok.

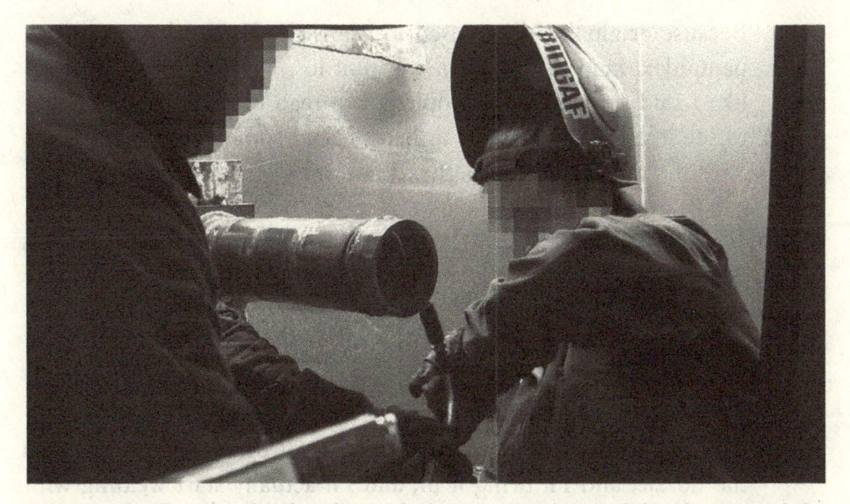

Figure 4.16. Tom put his hands on the gun as Stephanie did a dry run of the pipe weld so that he could rotate the gun for her: "If you take this, and rotate it." Note: My audio recorder occasionally got in the frame during recording. Tom's left hand appears just behind and right of the audio recorder in the foreground. His right hand was further down the gun and out of the frame.

In response to Tom's question, Stephanie confirmed that she understood the movement correctly, performing it simultaneously. After offering further explanation for using this method to weld around the pipe, Tom told Stephanie to try another dry run so that he could assess her entire movement around the pipe:

Excerpt 4.8 (continued)
TOM: Do one more dry run. I'll watch. [4s] Is that better?
STEPHANIE: Yeah.

Through his series of pumping questions and telling strategies, Tom provided support within Stephanie's ZPD. When she couldn't supply an answer, he constrained her possible responses or even provided an answer.

In addition, to help her feel a comfortable movement, Tom took hold of the gun and moved along with Stephanie. Discussing how he draws on a range of resources to meet the needs of students with various ways of learning, Ted addressed the topic of physically guiding a student through an action:

TED: Because originally all we used to use is the Hobart little pamphlets kind of. But then we went to a big book, and because like I said, we get some people got to read it. Some people got to see it. Some people got to hear it, and some people have to have me grab them by the hand and make them do it.

When I asked him to elaborate on physically guiding students' hands during a weld, he responded:

TED: [In] stick I will actually grab the rod. I will tell them right here. I say, "I'm going to grab the rod, and I'm going to run the bead from down here, and I'm going to run the bead. You keep doing what you're doing." Or I'll grab the end of the MIG gun. I'll grab the nozzle, and I'll bring it in, and I'll actually start welding with it just holding it by two fingers. . . . And I'll just weld with it like that, and then they'll get the feel of it, and they're like, "Oh wow, I'm too far out, or I'm too- Now I see. Now I can see how close I need to be."

Ted's comments point to the pedagogical utility of physically guiding students' actions, what Andersson, Öhman, and Garrison (2018) called denoting touch. This type of pedagogical touch, they wrote, "is characterized by intentions to handle learning content. Teachers use denoting touch to convey the desired body movement and help the students attend to certain qualities in order to develop different motor skills. It is about bringing the body into the right position to create a sense of the movement itself" (600). They differentiated this type of touch from security touch, which aims to prevent injuries, particularly in high-risk situations (599) and relational touch, which aims to generate "good relationships and atmosphere" (602). With both denoting touch and security touch, teachers help students learn "to correctly intuit a quality" (601), such as comfort, and to control their bodies. In certain situations, they wrote, "pointing with words or the use of metaphors will not always do the job" (601).

With physical guidance of a student's movement via touch, teachers can further students' learning in several ways. Discussing the use of touch when teaching voice, Thuma and Miranda (2020) pointed out that teachers can show students the "optimal position of body parts in relation to each other" (218), as Tom did when he helped Stephanie position her arm so that she could move continuously around the pipe. Another pedagogical use of touch, they wrote, is to draw attention to an area of the body, for example, one that is too tense (217), as Tom did when he touched Stephanie's wrist to draw attention to it. Thuma and Miranda (2020) also wrote that touch can "bring attention to the internal channels for learning," given that "[m]any people need a learning process to develop a more accurate body sense," including what is tense, what is bent, and what is tight (217). This last pedagogical use of touch equates to Tom's overall goal through this interaction with Stephanie; he wanted to help her develop her own sense of what is comfortable and what is not.

Research in the teaching of physical education, too, has explored the "politics of touch," examining the tension between teaching physical skills through touch and no-touch policies aimed at protecting students from abuse and harassment. Öhman (2017) found that physical education teachers, particularly men, were reluctant to touch students in order to help demonstrate movements. Nevertheless, Caldeborg, Maivorsdotter, and Öhman (2019) found that students found touch acceptable when they were learning new techniques or when they perceived the teacher to be keeping them from injury, contexts that align with Andersson, Öhman, and Garrison's (2018) denoting touch and security touch.

Conclusion

Part of learning how to be a welder is learning to perceive as a welder does. In scaffolding students' expert perception, welding teachers used cognitive scaffolding strategies, namely pumping questions, to diagnose whether students perceived—saw, heard, or felt—some phenomenon—and thus obtained students' current level of understanding. Their pumping questions often highlighted some phenomenon that the welding teacher would later imbue with meaning, or code, using explaining strategies. To further highlight visual phenomena in the shared space, teachers layered deictic gesticulations onto their verbal tutoring strategies, as when Tonya pointed to a specific location on Steve's weld or when Tom pointed to Stephanie's wrist.

With their pumping questions, teachers also made conversational space for students' contributions and thus facilitated greater interactional interactivity. Even closed pumping questions such as yes-no questions provided a conversational opening for students—an opening that they could fill with a more substantial response, as when Teresa used a pumping question to confirm Saul's interpretation of the machine settings: "It was a little cold, huh?" In response to this pumping question, Saul began with a minimal response, "Yeah," but then continued on to explain his setting preference: "And I- I tend to like to run right to the edge of-." While students—particularly ones less experienced than Saul—may not expand upon a minimal response, pumping questions at least afforded them a turn at talk and gave them an opportunity to elaborate upon their initial answer or to ask a question, as Stephanie did in excerpt 4.5. Given their role in facilitating diagnosis and generating interactivity, welding teachers might try to employ more pumping questions, even closed pumping questions.

Also, when teachers note a particular phenomenon and want to highlight it, they might ask broader, more open-ended pumping questions. Such questions could call upon students to identify the phenomenon of interest, for example, big drops of filler metal. An open-ended question, such as "What do you think you should notice at this point?" would alert the student to the fact that the teacher considers some phenomenon as worthy of attention. The student would then have the opportunity to analyze the environment—its sights, sounds, feel, and potentially, smell—and offer a response. Cases in which students point to some phenomenon other than the one the teacher had in mind could also lead to useful discussions, for example, about why that particular phenomenon, like the fluidity of the weld puddle or the consistency of the weld, is in fact normal or satisfactory and what might occur to change that status to less normal or less satisfactory. Such extended conversations could also allow teachers to further diagnose a student's level of understanding, enabling the teacher to better determine the type of the next intervention and the degree to which it would constrain the student's response.

Welding teachers might also employ other cognitive scaffolding strategies—ones not observed in this study's data—as they diagnose students' current understanding and help them develop expert perception. Other cognitive scaffolding strategies, such as prompting and forcing a choice, drive students toward a response but limit that response's degrees of freedom (see Cromley and Azevedo 2005, 97; Mackiewicz and Thompson 2018, 38). With a prompting strategy, a teacher sets up the student's response by providing a partial response or by leaving a blank for the student to fill in. For example, a teacher could use a prompt, such as "For spray transfer, the ratio of argon and CO_2 is, what?" With a forced choice strategy, the teacher provides options, such as, "For spray transfer, do we use an argon-CO_2 mixture of 90/10 or 75/25?" Such questions could help teachers determine students' knowledge and push them to participate.

Some welding teachers might consider prompting and forcing choice strategies like the examples above to be too much like test questions, generating situations in which students respond incorrectly, lose face, and become discouraged. But teachers can formulate prompts and forced choices so that they push students to think out loud about their responses. A teacher might formulate a prompt like this: "Think out loud about an answer to this question. A good way to fix this spatter problem is to do what with the voltage?" Or, a teacher might use a forced-choice strategy like this: "Ok. Think through your answer to this question out loud. Given the results we're getting here, do you think I'm going to tell you to turn the voltage up or down?" Such questions rein in students' possible responses but also create opportunities for teachers to determine students' understanding, including students' understanding of abstract concepts and their outward manifestations in weld results.

After a shared perception of a highlighted phenomenon, or intersubjectivity, has been established, the welding teacher might use pumping questions to prod a student's ability to code that phenomenon, as Teresa did when she asked Saul to assess the penetration of his weld (excerpt 4.3). This question made room for Saul's own explanation, that is, for coding as opposed to highlighting. Such pumping questions would push students further in their thinking, moving them from "what it is" to "why it is."

In addition to using pumping questions, teachers used the strategy of referring to a previous topic, connecting new information to old information and new tasks to familiar ones and thus building intersubjectivity. By referring to previous topics, teachers helped students bring to bear knowledge that they already possessed in order to employ it in a new situation. It's worth pointing out that students, participating actively in their own learning, also tied the task at hand to what they had learned before. Building off Tonya's pumping question that referred to a previous lesson, Sutton asked questions of his own that recalled what he had learned before:

Excerpt 4.9

TONYA: You might want to- You remember how I took your temperatures down when you were going up?

SUTTON: Mmhm.

TONYA: [Nods]

SUTTON: And move the wire feed down?

TONYA: Yeah. Anywhere from 180 to 220. And your voltage's right around 18.
SUTTON: Ok.

Tonya's pumping question pointed to a prior lesson that applied to the current task—turning down the voltage for a vertical-up weld. After a minimal response ("Mmhm"), Sutton continued his turn at talk to confirm that in this task, as in the previous one, he should lower the wire speed. His confirmation question indicated that he was indeed able to recollect what he had previously learned. Seeing that he remembered, Tonya built on his knowledge, redirecting him from wire speed to voltage. In response to Tonya, Sutton asked yet another question, this time confirming that he could use the same technique that he had before:

Excerpt 4.9 (continued)
TONYA: Just once you get- get the voltage on 18 and then adjust the wire for your [speed.
SUTTON: [And then do the tight G's again?
TONYA: Yeah.

By linking Sutton's new task to what he already knew, Tonya spurred interactivity from Sutton as he asked questions to confirm that the machine settings and technique that he had learned before would apply in the current task as well.

Expert perception develops slowly, over the course of semesters in school and years on the job. In welding classes, students learn where they should put their attention as teachers differentiate figure from ground to highlight phenomena critical to generating a good weld. Slowly, students learn to code those phenomena as teachers explain their meaning (or their multiple potential meanings). In short, in scaffolding students' development of expert perception, teachers moved students forward on the path toward membership in welding's community of practice. In the next chapter, I examine how teachers helped to develop and maintain students' motivation as they traveled this path.

Chapter 5

Developing and Maintaining Students' Motivation

My SMAW teacher assured me that everyone needs several class periods to complete a horizontal open-root weld. Indeed, she said, some students never complete the task. After four class sessions spent on the weld, I was beginning to think that I was going to be one of those destined for failure, even though I had been coming to class a half hour early to get my machine and tools set up so that I could devote the entire class period to more (endlessly sad) attempts. While grabbing yet another stack of coupons from the bin, I encountered a teacher, Tarek, who taught the class that let out at 5:00 p.m. Because I had been coming early, we'd run into each other a few times, chatting as he prepared to head home and as I prepared for another night of frustration. When I told Tarek that I'd been struggling, he too told me that particular weld is really hard for everyone and even assured me that when I go out into industry, I likely won't have to do it. His sympathy helped, but I still wanted to join the ranks of those who have run at least one acceptable horizontal open-root weld.

As Teresa pointed out to Sophia (see excerpts I.3, 3.12, and 4.6), an open-root weld in SMAW involves waiting for the keyhole to open up and then moving the electrode across the joint at an even pace, being careful to maintain the keyhole but not allowing it to become so large that it's no longer possible to weld the two pieces of steel together. In the previous four class periods, I'd blown out more of those welds than I could count. Or, I'd not managed to wait for the keyhole and thus not achieved weld penetration. The travel speed across the weld—not too fast, not too slow—seemed impossible to achieve and maintain. Throughout my attempts, my SMAW teacher, like Tarek, offered words of encouragement: "You're getting it." "It's really hard." "Keep going." I was frustrated with my lack of control over my body and my inability to see and hear the keyhole, but my teacher's words motivated me. I kept going.

Although I often felt like I was the only frustrated person in the class, I know I was not. When I asked him about students' frustration with the pace of their progress, Tom told me about a metaphor he uses to talk about welding:

TOM: And it- it's just- a lot of people will say welding is a skill. And then the analogy I use is it's more of an art form. Like, when you look at all the little intricacies- And sort of how that skill develops and- and so then it's, you know, referring to famous artists, it's- you know, they didn't learn to do that in six months.

By comparing welders to artists, Tom helped students to consider the hours of practice needed to advance their competence:

TOM: And [I say this] to kind of help ease them when they start coming up against those kind of learning walls where they need to kind of break through those to progress. . . . They start getting frustrated, and I'm just kind of reminding them of that.

The "learning walls" that Tom spoke of require students like me—those of us frustrated by our repeated failures—to be persistent in our efforts. And while I think that tenacity is a strong component of my personality, I know that what kept me in my booth practicing the same weld (such as the infamous horizontal open-root weld) over and over was the encouragement of my teachers. In fact, when I asked Teresa about how she got started in welding and whether she had received encouragement, she told me that it was, ironically, a jerk of a boss who had encouraged her to pursue welding, perhaps his one redeeming act. At the time, she was working at an auto shop, having graduated from an automotive program:

TERESA: I got a job and that was my first . . . Because I have a bachelor's degree in anthropology with a minor in women's studies. So I had this idea that everything's equal. You know, I don't know what I was thinking, but it was a big slap in the face when I realized that sexism's really alive and obvious. It's

> not even just subtle. So I had a real jerk of a boss. . . . Words came out of his mouth that I just couldn't believe. I was just shocked. But I learned a lot, and I got what I was lacking in that first job was confidence because I was trying so hard. But I got an opportunity to go back to school for welding for a year. And I kind of decided I didn't like my boss, and I had asked for the raise he had promised me, and he didn't give it to me. And so I decided I'm leaving, and I learned a little bit of welding there. I started with exhaust work. And that was actually one thing he was encouraging me with.

Teresa's story underscores the power of motivation, a critical affective component of learning: It can make a difference even when it came from the mouth of a sexist ass.

As Vibulphol (2016) summed up, "Without motivation, learners may not start the act of learning at all and for those who have started to learn, they may not be able to maintain their learning once experiencing hardship in the process (64; see also, Dörnyei 2001; Gardner 2007; Palmer 2009). Indeed, research has suggested that students who lack motivation are less attentive in class (Goetz and Hall 2014) and are less likely to achieve the intended learning outcomes (Pekrun et al. 2010). Motivated students, it seems, are more likely to attempt challenging tasks and to take academic risks (Turner and Meyer 2004). Describing students who are motivated to learn, Brophy (1983) wrote, "[They] will not necessarily find classroom tasks intensely pleasurable or exciting, but they will take them seriously, find them meaningful and worthwhile, and try to get the intended benefit from them" (200). In sum, much research has suggested that teachers' behaviors can positively influence students' affect, specifically, their motivation, and that motivation is critically important to students' persistence in learning (e.g., Bolkan and Griffin 2018; Mazer 2012; Moskovsky, Alrabai, and Paoli 2013; Papi and Abdollahzadeh 2012).

This chapter examines the ways that welding teachers used five motivational scaffolding strategies (see table 5.1) to foster student persistence throughout scaffolded learning. In particular, motivational scaffolding strategies appear to positively influence students' extrinsic motivation. Extrinsic motivation refers to external rewards such as recognition that can help generate tenacity.[1] In contrast, intrinsic motivation refers to personal satisfactions, including curiosity. In my experience, welding students are often intrinsically motivated, driven by a personal interest in

Table 5.1. The motivational scaffolding strategies found in this study's data

Strategy	Definition
Giving sympathy	Teacher acknowledges that the task is difficult for the student.
Being optimistic	Teacher conveys positivity by asserting a student's future ability to succeed in a task.
Praising	Teacher points to student's achievement with positive evaluation. Praise can be formulaic or nonformulaic.
Showing concern	Teacher builds rapport with students by demonstrating that they care.
Using humor	Teacher kids around, tells jokes, or tells amusing stories.

acquiring competence and entering a community of practice. The extrinsic motivation of motivational scaffolding strategies, though, can help get a student through those horizontal-open-root-esque challenges that might otherwise make that student pack it in for the day.

Giving Sympathy

When welding teachers gave sympathy, they acknowledged the difficulty of the task at hand, as both Tarek and my SMAW teacher did. Prior research has shown that more difficult tasks decrease motivation, a finding I have personally experienced (Glynn 1994; Horn and Maxwell 1983; Swift and Peterson 2018). In giving sympathy, teachers helped students see that their struggle to accomplish some tasks competently was understandable and, more important, expected: most tasks in welding are difficult to do at first, and some tasks, such as welding in the overhead position, even experienced welders find difficult to do well. I would argue, too, that when teachers acknowledged the difficulty of acquiring a skill or understanding a concept, they helped to create intersubjectivity—that shared understanding of the task at hand that is a critical component of scaffolded learning.

Welding around a pipe is difficult and thus typically taught at the end of the semester, if at all. After Tom had demonstrated welding around a pipe and had helped Stephanie find a more comfortable body position throughout the weld (see excerpts 3.4 and 4.8), he ended the interaction by noting its difficulty, saying that "it's a lot of moving parts":

Excerpt 5.1

TOM:	You have a tendency to want to push it out at the bottom. You can spin that, to where you're actually turning it so that the gun is away from you, to get your drag angle at the bottom.
STEPHANIE:	Ok.
TOM:	It's a lot of moving parts. You got it.
STEPHANIE:	Ok.

This show of sympathy acknowledged that welding around a pipe means coordinating moving arms, torso, and legs together to create a flowing movement. Indeed, as he used this giving sympathy strategy, Tom moved his hands up and down vertically in a flurried motion (figure 5.1), a metaphoric gesticulation that further conveyed the task's complexity—its coordinated body movements that at first seem quite chaotic.

Tom followed up his sympathy with optimism, another motivational scaffolding strategy: "You got it." With it, he predicted that Stephanie, who had just performed a dry run of the movement, would carry out the weld successfully when she tried it again.

Figure 5.1. Tom used a metaphoric gesticulation, moving his hands up and down vertically to indicate "a lot of moving parts."

Similarly, when Suzanne told Teresa about the difficulty she was having with feeding the filler rod through the fingers of her left hand without moving the rod with her right hand, Teresa quickly pointed out that the action was indeed awkward, especially with bulky welding gloves on:

Excerpt 5.2

TERESA: That takes a little practice. I mean you can even take a rod home and-

SUZANNE: [Right. [unclear] with gloves. [unclear]

TERESA: [and just sit there-

TERESA: Yeah. It's really weird with the gloves, um.

Teresa expressed sympathy first when she noted that the technique of smoothly feeding the filler rod through the fingers of one hand "takes a little practice." Her use of the understater[2] "a little" downgraded her claim somewhat, allowing Teresa to both acknowledge the difficulty of the task and still convey that competency was a quite achievable goal. Teresa elaborated with another show of sympathy when she further acknowledged that, yes, unwieldy welding gloves make the action of moving a filler rod through one's fingers particularly challenging.[3] Both Teresa's acknowledgment that the action "takes a little practice" and her recognition that "it's really weird with the gloves" ratified the frustration that Suzanne felt and supported her suggestion that Suzanne take a filler rod home to practice.

The showing sympathy strategy promoted students' motivation in that it made them feel that they were not alone in struggling with the task because the task was indeed difficult. In addition, with this strategy, teachers helped build a shared understanding of the task. Such intersubjectivity made scaffolded teaching and the learning that resulted possible. With intersubjectivity, teachers and students jointly developed and redeveloped a shared understanding of what was being taught and learned moment to moment. Through it, teachers and students shared a context that both framed and was framed by their contributions (both verbal and visual) to their interactions.

Being Optimistic

When welding teachers used the strategy of showing optimism, they conveyed their belief that a student would complete a given task successfully.

As Furnham (1997) put it, "Optimists, by definition, emphasize either or both favorable aspects of situations, actions, and events in the current world as well as believing in the best possible outcomes in the future world" (198). Research examining the effects of teachers' individual optimistic statements is lacking, but one study of note is Vehkakoski's (2019) conversation analysis of primary-school special-education teachers' use of optimism to counter students' expressions of failure expectation, such as "I can't get it" and "I don't know" (3) and to boost students' own optimism about their learning. Examining 52 discourse sequences in two teachers' talk, Vehkakoski found that the teachers countered failure-expectation expressions most often (52% of occurrences) with "inversions" such as, "Yes, you can" (6). With inversion expressions, the teachers showed "confidence in the students' coping potential" and in turn minimized "the existence of learning problems" (6). Vehkakoski also found several other strategies that promoted students' optimism, including giving examples of peers' success,[4] and referring back to a student's earlier successful performance. Although studies like Vehkakoski's are lacking, research in educational psychology has quite thoroughly examined the role of teachers' expectations on students' performance (e.g., Luthans, Luthans, and Jensen 2012; Tetzner and Becker 2017; Urhahne 2015). Such research has shown a strong connection between teachers' expectations for students' success and students' motivation (Jussim 1989; Jussim and Harber 2005; Phan 2016).

In addition, I would argue that optimistic statements foregrounded teachers' macrolevel scaffolding, that is, the sequencing of lessons, tasks, and tests throughout the semester and, as well, throughout the program. When welding teachers used optimism, they tended to speak of the way that building on prior knowledge—whether from practice with the same task, practice with related tasks, classroom lecture, or textbook content—would make future tasks easier or more understandable. For example, in closing her interaction with Stewart, Tonya asked whether he had reread the chapters in his textbook that covered oxyacetylene welding. His answer prompted a suggestion, an explanation, and finally, a statement of optimism from Tonya:

Excerpt 5.3

TONYA: Ok? Did you reread your oxy chapter?

STEWART: Uh, not since last time.

TONYA: Yeah, that's a- It would be a good idea for you to go over those chapters.

Tonya's metaphoric gesticulation as she suggested that Stewart "go over" the chapters on oxyacetylene helped convey the idea that the types of input Stewart was getting—the learning that came from reading and processing his textbook in addition to the learning that stemmed from his practice time in the lab—were iterative and mutually strengthening (figures 5.2 and 5.3). After Tonya made this suggestion, she used an explaining strategy to

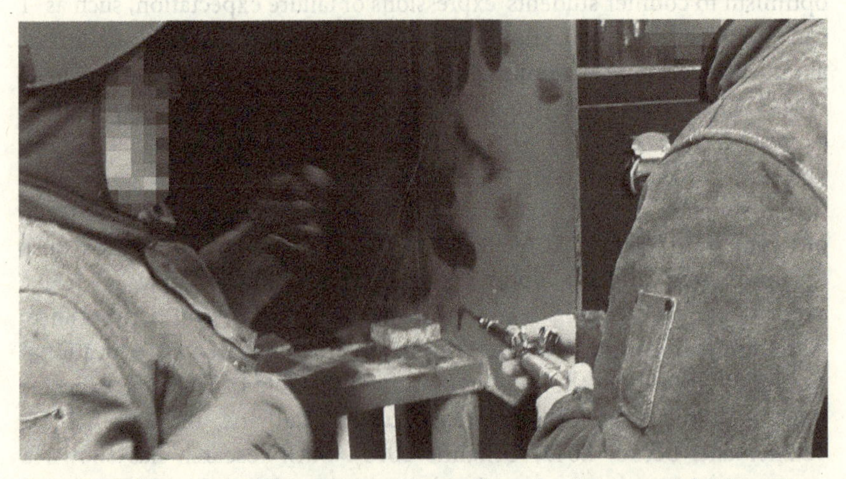

Figure 5.2. Tonya rolled her hands one over the other in a metaphoric gesticulation as she suggested that Stewart reread, or "go over," the textbook's chapters on oxyacetylene welding.

Figure 5.3. Tonya continued to roll her hands one over the other in a metaphoric gesticulation as she suggested that Stewart reread, or "go over," the textbook's chapters on oxyacetylene welding.

provide her reasoning. As she did, she pointed, using an abstract deictic gesticulation, reinforcing the idea that the conceptual knowledge would eventually manifest concretely in Stewart's welds (figure 5.4):

Excerpt 5.3 (continued)

TONYA: Because they'll make a lot more sense now that you're actually applying it.

In this case, then, the teacher's optimism explicitly connected the textbook's contents to the practice of welding, making clear to Stewart that the embodied knowledge that he wanted to develop hinged on his ability to develop conceptual knowledge simultaneously.

In a similar manner, Teresa connected Sophia's ability to carry out more advanced welds with her practice with the horizontal open-root weld. Using the showing optimism strategy, she asserted that those out-of-position welds would be easier because Sophia had practiced a (dreaded) horizontal open-root joint. As she named the two future out-of-position welds, she illustrated them with iconic gesticulations (figures 5.5 and 5.6 and figures 5.7 and 5.8). Then, stating the relevance of her current task to those welds, Teresa pointed to and even tapped the bottom of the horizontal open-root joint in front of them (figure 5.9):

Figure 5.4. Tonya used an abstract pointing gesticulation, indicating the weld that she had just completed and the space where Stewart would carry out the same process.

Excerpt 5.4

TERESA: Once you get it though we're going to do the same root in the vertical-up position and the overhead position and those positions will go a lot easier because of the time you're going to spend on this horizontal one.

SOPHIA: Ok.

Figure 5.5. Teresa raised her hand up in part of an iconic gesticulation to indicate the action of performing a vertical-up weld.

Figure 5.6. Teresa lowered her hand from its raised position, completing an iconic gesticulation to indicate the action of performing a vertical-up weld.

Figure 5.7. Teresa moved her hand toward her head, beginning an iconic gesticulation to indicate the action of performing an overhead weld.

Figure 5.8. Teresa then moved her hand out in front of her, using her elbow as the fulcrum and completing her iconic gesticulation to indicate the action of performing an overhead weld.

TERESA: Ok?
SOPHIA: All right. Sounds good.

Using these three gesticulations one after the other further tied them together: Welders must be able to move the gun steadily on a horizontal

Figure 5.9. Teresa used a concrete deictic gesticulation to point to the horizontal weld in front of them, tapping the lower coupon after she pointed.

axis before they can move on to a vertical axis and finally to a weld in the most uncomfortable position of all—overhead.

Before ending the interaction, Teresa used another show of optimism. On two occasions, Teresa had already warned Sophia that generating successful horizontal open-root welds would take several class periods. Indeed, on one of those occasions, she noted that Sophia should just "be ok" with this fact. She also pointed out that if Sophia were to ask other students who had already accomplished the challenging weld, they might tell her "it took them even longer." Even so, Teresa optimistically declared that Sophia might achieve the desired outcome even faster:

Excerpt 5.4 (continued)

TERESA: But maybe you'll get it done in one. That would be awesome, huh?

SOPHIA: Right?

TERESA: All right. So try a couple and I'll check in on you and see how it's going.

SOPHIA: Ok. Sounds good. All right.

There didn't seem to be any particular reason to think that Sophia would advance through this weld position any faster than other students—at least no reason that I could tell. Rather, Teresa's optimism seemed intended to remind Sophia that once in a while, for whatever reason, a student simply gets the knack of a given task easily. On these happy (and, I think, rare) occasions, the student can move on to the next task early.

Excerpts 5.3 and 5.4 exemplify how welding teachers—when they did show optimism—used that optimism to tie old to new knowledge or to tie conceptual knowledge to applied knowledge. In this, their motivation not only provided affectual support—encouragement—but it also helped students understand their learning process.

Using Praise

Of all the motivational scaffolding strategies that appeared in welding teachers' talk, prior research, it seems, has examined praise the most. That research has shown that praise—expressing appreciation of a student's achievement—can often have strong positive effects on students' motivation and, in turn, their learning. For example, as Lepper et al. (1993) pointed out, praise can promote students' self-efficacy, which is their belief in their ability to act in order to achieve the outcomes they want (see also, Bandura 1977, 1986, 1997). Praise can also increase students' interest in a task, given that students tend to be more interested when someone, especially someone with expertise, shows appreciation for their work (Hidi and Boscolo 2006; Margolis 2005). According to O'Leary and O'Leary (1977), praise is most effective when it is not only sincere and contingent upon a successful performance but also specific (see also, Brophy 1983).

One way to analyze the specificity of praise is to distinguish between praise that is prepatterned, or formulaic, and praise that is novel, or nonformulaic (Mackiewicz and Thompson 2018, 134). Formulaic praise follows a syntactic pattern, and it often contains the common and general adjectives "good" and "great." Syntactic formulae from this study's data included the following:

- noun phrase + "is" + adjective phrase ("You're weld here is good")
- adjective + noun phrase ("Nice job")
- "I" + ("really") + "like" + noun phrase ("I like your tool setup") (see Mackiewicz 2006).

Nonformulaic praise, in contrast, was more context specific and thus indicated that it had been generated for a particular individual. Because it clearly applied to a specific accomplishment, it might also have been more likely to reinforce positive behaviors (Brophy 1983; O'Leary and O'Leary 1977). In this section, I discuss how welding teachers used praise—both formulaic and nonformulaic—to reinforce positive behaviors, to balance negative evaluations, to close down interactions, and to point to students' progress.

Reinforcing Positive Behaviors

Prior research has shown that praise has reinforcing effects in academic contexts (Brophy 1981, 1983; Hall, Lund, and Jackson 1968; McLaughlin 1982), making salient the behavior that the teacher wants to emphasize and fortify. After watching Simon complete a vertical-up weld in SMAW, Tonya used nonformulaic, specific praise for Simon's putting into practice the lesson that Tonya had given him earlier:

Excerpt 5.5
Simon: Ok.
Tonya: You know you'll- you'll find you'll have a lot more puddle control- That- But you did exactly what I told you to do. Soon as that puddle mushrooms out you're with it.

Tonya praised Simon for consistently whipping the 6010 electrode forward when the puddle formed. With her praise, Tonya reinforced Simon's behavior.

Welding teachers also used praise on a moment-to-moment basis during their guidance of students through a task, encouraging students to continue a behavior they had just performed. For example, Ted used formulaic, general praise ("That's good") as he directed Sawyer through an overhead weld in SMAW:

Excerpt 5.6
Ted: Faster. Faster. Faster. That's good. Stay moving with it. High. There you go.

While it seems clear that nonformulaic praise tends to be more informative in that it is individualized, in cases in which the teacher must respond in the moment, as when Ted observed and commented upon Sawyer's overhead welding, formulaic praise afforded the ability to comment quickly while students were performing actions. And because formulaic praise is prepatterned and likely retrieved as one linguistic unit, it is perhaps easier to generate in the moment, enabling teachers to keep students on track.

Balancing Negative Evaluations

Teachers' praise helped to balance teachers' negative evaluations, or criticisms, of areas in which students needed improvement. Negative evaluations, though expected in scaffolded teaching, are nevertheless threats to positive face (Brown and Levinson 1987). Just as praise bolsters positive face—the want to be appreciated by others—criticism detracts from positive face. One more polite way to deliver a negative evaluation, then, is to balance it with praise. With such a balance, teachers can articulate areas that need improvement while mitigating discouragement and maintaining motivation.

Tonya had returned to Steve's booth to check on Steve's progress with vertical-down lap weld using GMAW. As the two looked at Steve's work, Tonya used her expert sight to interpret what had happened when Steve had performed the weld. After reading the weld, Tonya began her assessment with praise that sat on the boundary between formulaicity and nonformulaicity:

Excerpt 5.6

TONYA: You're starting out perfect. That's beautiful right there.
STEVE: Mmhm.
TONYA: But then as you come along you're moving it back in and pointing to that upper leg.
STEVE: Mmhm.
TONYA: And you've got to point that back- the back end of the joint. That's why the puddle's getting ahead of you and it's rolling over.

Her praise, while quite formulaic in both syntax and lexicon, was strong, certainly conveying to Steve that he had achieved excellent results for at least a few moments during his performance of the weld. Accompanying her praise was a concrete deictic gesticulation, which she used to highlight the section of the weld that was both "perfect" and "beautiful" (figure 5.10).

Her concrete deictic gesture continued, however, as she quickly switched from this praise to the bad news—where Steve had gone wrong

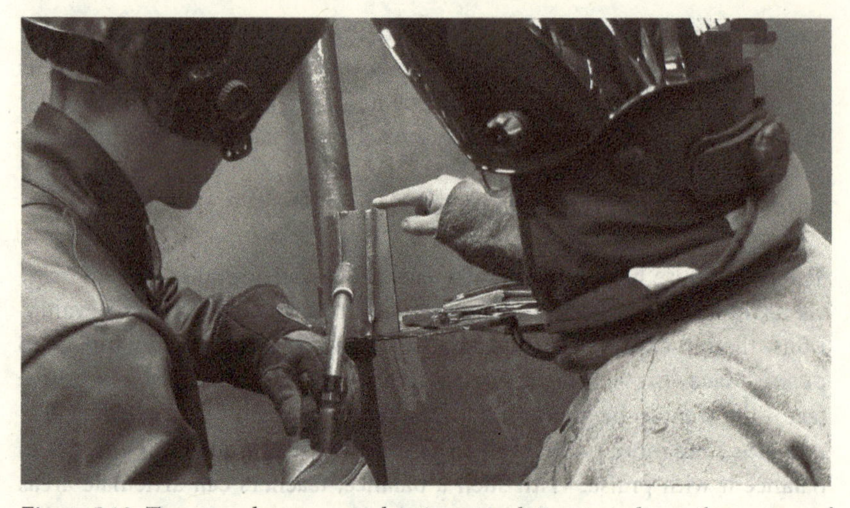

Figure 5.10. Tonya used a concrete deictic gesticulation to indicate the section of Steve's weld that was "perfect" and "beautiful."

with the work angle: "But then you come along- you're moving it back in and pointing to that upper leg." The negative evaluation contained in her reading set up what came next—a suggestion to change his work angle and an explanation of the problem that had manifested thus far. In this, her praise buffered harder news to come.

These times when teachers used their expert sight to read students' welds provided opportunities for teachers to offer negative evaluations and advice for improvement. In turn, they offered opportunities for teachers to use praise to mitigate these potential motivation drains. Excerpt 5.7 exemplifies praise used after a negative evaluation to encourage a student to persist. Teresa returned to Susan's booth to check on Susan's progress on a horizontal weld in SMAW. Earlier, Teresa had visited and had told Susan to practice a straight dragging motion. On this second visit, Teresa told Susan how her practice would progress; Susan would continue to practice a steady drag across the weld. Then, upon performing that successfully, she would move on to a weave motion. In this, Teresa explicitly pointed to the macroscaffolding of the class—how Susan would progress from one skill to the next challenge. Then, Teresa articulated her assessment, first negative evaluating Susan's travel speed in one section of the weld, using a concrete deictic gesticulation (much like Tonya in excerpt 5.6) to indicate the exact location on the weld that she meant (figure 5.11).

Figure 5.11. Teresa used a concrete deictic gesticulation to indicate the place on the weld where Susan had moved too slow with the electrode.

But Teresa immediately followed up her negative evaluation, itself softened with the understater "a little," with hedged praise: "but most of it looks good":

Excerpt 5.7

TERESA: So I'm going to have you do a steady drag for the first couple and then we're going to move up to that little W motion to help flatten it out a little bit.

SUSAN: Ok.

TERESA: When you go to that little W motion it'll help flatten it out here. [So you went a little slow right here but most of it

SUSAN: [Yeah.

TERESA: [looks good.

SUSAN: [Ok.

Teresa's use of a downgrader ("a little") and a hedge ("most of it") exemplified how teachers calibrated their evaluations, balancing the need for honest assessment with the need to attend to students' affect, including their level of frustration. Indeed, as she moved from critique to praise, Teresa moved her hand away from the weld and back to her body. For her part, Susan brought the weld closer to her face to examine it closely (figure 5.12), trying to read it—to perceive it with expertise—as Teresa just had. Like Tonya, Teresa balanced evaluation that was necessary for developing students' ability to connect their results to their performance with sustaining their motivation to continue to their efforts. This strategy seemed to have a positive effect, as Simon's interview comment suggested:

SIMON: I've never welded before besides this summer. And I had never done sixty ten [a 6010 electrode] before. And when I first started, my bead was really bad. And she [Tonya] told me, "It's a good width, but you really need to improve the length between each bead." And she'll say things like that. Like, when I did MIG horizontal. You know, she told me, she says, "Your bead's fine and your travel speed is fine, but you have to improve your travel angle." She always adds something nice to say about it before she tells you what you need to fix.

Closing Down Interactions

As excerpts such as 5.1, 5.4, and 5.7 suggest, welding teachers sometimes closed down interactions with praise as a final boost of encouragement before leaving students to practice on their own. Tom had visited Sebas-

Figure 5.12. Teresa moved her hand back toward her body and away from the weld when she used an understated praising strategy: "but most of it looks good."

tian's booth to check on his progress and evaluate his welds so far. After taking a look, Tom adjusted the settings on Sebastian's machine and told him that he would watch him weld in order to assess his technique. After Sebastian ran a bead, Tom pointed out that the popping they heard meant that Sebastian would have to adjust his machine settings further to get better results. After this assessment of the situation, Tom evaluated Sebastian's results, the weld Sebastian had just completed, with formulaic praise: "The speed back here looks good." Then, Tom used formulaic praise ("Nice job") to indicate that his interaction with Sebastian, with its machine adjustments and observation, had run its course:

Excerpt 5.8

Tom:	Yeah. The speed back here looks good.
Sebastian:	Ok.
Tom:	So. Nice job.

Tom's general "Nice job" ended the progress check on a positive note that could contribute to Sebastian's motivation to persist in troubleshooting the problem after Tom had left the booth (figure 5.13).

Teachers also used closing praise to balance a final negative evaluation and telling and suggesting strategies pointing to what the student

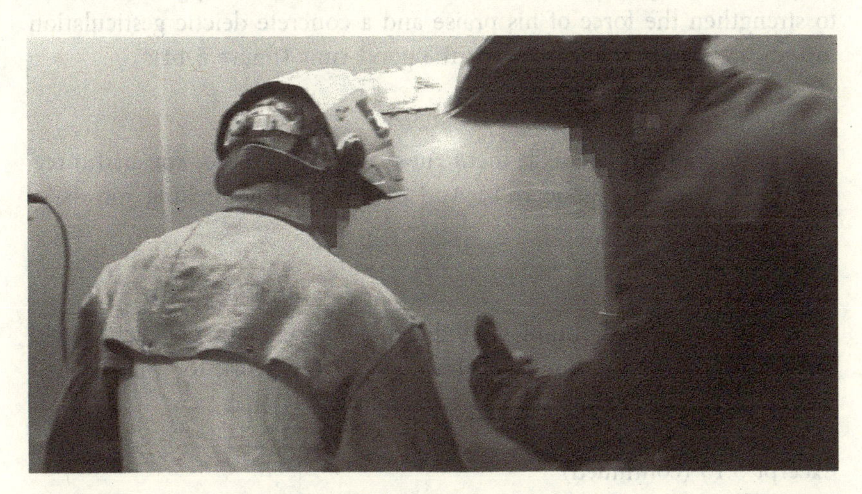

Figure 5.13. Tom performed a thumbs-up gesture as he used a nonformulaic praising strategy ("But the speed back here looked good") and before he finished with a formulaic praising strategy ("So, nice job"). Thumbs-up is an emblem gesture.

should practice. In winding up her interaction with Shawn, for example, Tonya used formulaic ("But that was really good") and nonformulaic praise ("That was what I was looking for there") before she reiterated the flaw that she had identified in Shawn's weld a few minutes before:

Excerpt 5.9

TONYA: Carry on. But that was really good. That was what I was looking for was there. You're just a little too much in the middle.

SHAWN: Ok.

In her good news-bad news combination, Tonya (like Teresa in excerpt 5.7) used the downgrader "a little." Indeed, her praise included an upgrader,[5] "really," that further calibrated the force of her overall assessment.

POINTING OUT STUDENTS' PROGRESS

Scaffolded learning involves assessing students' progression toward competence in order to maintain a level of challenge within their ZPD. With praise, teachers marked students' advancement on the path toward embodied knowledge. Ted, for example pointed to Sawyer's improvement when he praised the quality of several weld runs, using the upgrader "really" to strengthen the force of his praise and a concrete deictic gesticulation to indicate clearly the location of the good runs (figure 5.14):

Excerpt 5.10

TED: Yeah, now look at them runs you got in there. You had three four really good runs. And that's what I was telling you that, you know?

SAWYER: Yeah.

After this praise, Ted continued on, this time offering a sort of backhanded praise to Sawyer. This praise evaluated his latest work to what he had accomplished or, rather, failed to accomplish previously:

Excerpt 5.10 (continued)

TED: That looks 100 times better than what you did already, right? All right. Let's move on. Keep going.

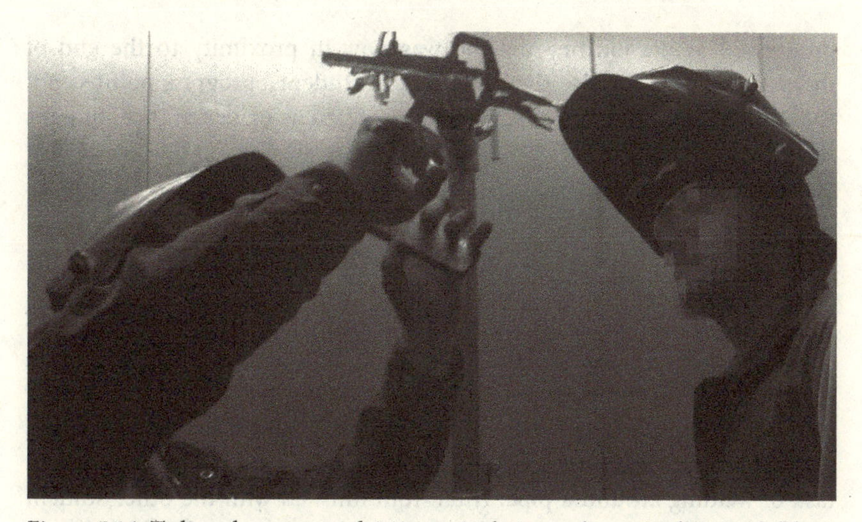

Figure 5.14. Ted used a concrete deictic gesticulation as he praised Sawyer's "three four really good runs."

While this praise gave positive face while taking it away from prior results, it had the benefit of making clear Sawyer's learning, his developing skill in the overhead position. Combined with subsequent directives to "move on" and "keep going," Ted indicated that Sawyer would make similar strides if he stayed motivated and kept practicing. Particularly in relation to overhead welding, Ted's praise of Sawyer's progress seemed well-timed. Overhead welding tires a person out. In response to a post on the American Welding Society forum about practicing overhead SMAW, TimGary (2002) described the challenge: "It's hard to maintain the proper distance from the end of the electrode to the workpiece when your arms are getting tired from holding them overhead. The proper distance to maintain is equal to the diameter of your electrode, or less. If you get tired and start 'long-arcing,' the bead will get sloppy and porosity will result."

This closing praise from Ted, as well as his telling strategies (e.g., "keep going"), employed a conventional metaphor—that learning is a journey along a path (see Berendt 2008; Cameron 2003; Jin and Cortazzi 2011). Used in scaffolding learning in welding, the physical act of traveling along a path becomes the source domain for the target domain of students' development of embodied knowledge. Teresa's praise in excerpt 4.3, "You're really close to having it" similarly conveyed that Saul had so progressed in his journey that he was now in proximity to the end of the path. In short, praise that pointed out students' progress pointed out the scaffolded learning that was going on acted as evidence that students should persist in their efforts.

Showing Concern

When teachers showed concern for students, they directly attended to students' affect, revealing with their words and actions that they cared about students' well-being and their learning. One instance of this strategy occurred when Tom appeared at Stephanie's booth after she had found him in the lab earlier to tell him that she was having difficulty with the task of welding around a pipe. After Tom finished with the other student he had been helping, he arrived at her booth as requested, asking her to explain the problem she was having:

Excerpt 5.11

TOM: All right, Stephanie, what's going on?

In response, Stephanie explained the trouble she was having with positioning her body and the gun so that she could maintain consistency throughout the weld as she moved from the top of the pipe to the bottom (see excerpts 3.7 and 4.8). Tom's question showed that he cared about her understanding and her potential frustration with this difficult task and that he was ready to help.

Tonya used questions that showed concern by offering Steve a chance to indicate whether he needed further clarification of the problems generated by laying more weld than needed:

Excerpt 5.12

TONYA: We want three-sixteenths is all the amount of weld you want. Over welding can get us in trouble. Over welding can cause things to break. Too much heat input into the steel. [2s] All right? Got it? Just point into that corner and drag it straight down.

STEVE: Ok. Thank you.

Although Tonya's questions could be interpreted as rhetorical, they did in fact offer Steve an opening to request further clarification. With these showing-concern questions, Tonya checked in with Steve to determine whether he had followed her explanation. As the analyses in chapters 3 and 4 have shown, welding teachers commonly used explanations to give reasons for their advice and to code the meaning phenomena in the shared space. However, in this study's data, teachers didn't typically stop to ask whether students had understood the explanations. It may be the case that welding teachers don't expect students to want or to benefit from long explications of conceptual knowledge while they are in the lab, as opposed to the classroom.

Using Humor

Welding isn't all seriousness all the time. Once in a while, teachers and students tell jokes. Once, when I returned from a bathroom break, my GMAW teacher and my welding buddy, Ryan, were telling jokes over the Lincoln welding machine parked in my booth. Before I had left for the bathroom, my teacher had demonstrated for me how to install a new spool of wire in the machine. (He had to do this several more times before I finally memorized all the steps.) When I returned, he was still standing by the machine, and Ryan had taken a break to join him. My teacher started a story about two hunters who are walking through the woods and find a deep hole in the ground. They wonder just how deep the hole is because it looks really, really deep. So, they proceed to throw a bunch of stuff down into it in order to gauge the depth, until they finally decide they need something really big to truly test its depths. So one of them walks away and soon comes back with an anvil. They promptly slide it into the mouth of the hole. Just then, they hear the sound of hooves coming toward them. It's a goat, and the goat jumps right down into the hole too. The hunters peer into the hole in amazement. Then, a farmer approaches. He's yelling "Becky! Becky!" He asks the hunters, "Hey, did you guys see a goat?" And the hunters say, "Yes! It's the damnedest thing. A goat just sped past us and jumped in this hole!" And the farmer says, "That's impossible. I had him tied to an anvil."

As we laughed, I realized that I had something to add to this comedic interlude. After welding class, I had been watching old episodes of Jerry Seinfeld's *Comedians in Cars Getting Coffee* on Netflix. (I usually couldn't fall right asleep after class, even though I got home late and I was exhausted.) My periodic viewing of Seinfeld's show had gifted me with a joke to tell—a Polish joke.[6] I told them about the Polish pilot and copilot who have been forced to land their plane because of bad weather. They land safely, but the plane ends up going off the runway and through the fence that surrounds the airport. It winds up in a field. The copilot turns to the pilot and says, "Damn, that was a short runway!" And the pilot responds, "Yeah, it sure was wide though." My joke got a laugh—an even bigger one than the goat joke received. I felt like I had joined the club.

The joke-telling exchange that night made clear to me humor's potential to build rapport. Fine (1987) wrote that humor can help create an idioculture, "a system of knowledge, beliefs, behaviors, and customs shared by members of an interacting group" (125). Through humor, including jokes, people make salient references that they share and thus make salient their shared identity as well (Fine and De Soucey 2005, 3–4).[7] Other researchers, too, have explored how humor binds certain groups together in that the humor depends on shared knowledge and, sometimes, shared challenges. For example, research on coping mechanisms has examined the usefulness of humor as a strategy for medical students and professionals (e.g., Madhyastha, Latha, and Kamath 2014). Some of this research has focused in particular on gallows humor (e.g., Piemonte 2015; Rowe and Regehr 2010).

The joke that I told in response to my teacher's goat joke did not rely on shared knowledge of welding. My sense of connectedness and in-group membership stemmed instead from my ratified participation in the exchange itself; I took the risk of taking the conversational floor, and my audience ratified my right to have the floor through their attentive listening and, after, through their laughter. As Fine and De Soucey (2005) wrote, "[C]o-presence is not sufficient for joking. It is interactive, and participation is required for it to succeed" (3). My success in this joke-telling episode made me feel like I belonged to this community of skilled-trades people.

Humor has other benefits. Substantial research, including research on people dealing with serious medical conditions (Boerner, Joseph, and Murphy 2017; Karami, Kahrazei, and Arab 2018) has shown that humor helps relieve tension and anxiety (e.g., Kuiper and Martin 1998) and mitigates stress (e.g., Abel 1998). Abel (2002) wrote that humor can "afford the opportunity for exploring cognitive alternatives in response to stressful situations and reducing the negative affective consequences of a real or perceived threat" (366). In a study of undergraduate students enrolled in a psychology course, Abel (2002) found that the students who rated highly on a test of sense of humor were more likely to use positive coping strategies and to reappraise stressful situations (376). In other words, humor helps us cope and thus makes us healthier.

Teachers' use of humor can generate important benefits to student learning. Studying 94 research methods and statistics students, Garner (2006) found that teachers' humor "can have a positive impact on content retention" (179; see also, Pollio and Humphreys 1996) and on students' perception of their learning. Korobkin (1989), for example, found that college students perceive that their learning is enhanced with appropriate humor. Teachers' humor can also generate student enjoyment of learning (e.g., Berk 1996; Brown and Tomlin 1996; Bryant, Comisky, and Zillman 1997). In addition, teachers' use of humor can generate and sustain student interest (Dodge and Rossett 1982; Machlev and Karlin 2017). And, as Bolkan and Griffin (2018) contended, based on their study of 281 communication-studies students, "instructional interventions" such as humor that capture students' attention are "just as important for student success" as other motivation-generating interventions (281). If students do not pay attention, they wrote, they cannot process new material and thus cannot make "long-term changes in their base of knowledge" (281). In other words, humor's ability to capture students' attention helps to create an environment that fosters learning.

Most humor in the welding lab was not the memorized-joke variety, occurring in downtime breaks. Welding teachers far more commonly used unplanned, off-the-cuff types of humor, such as gentle teasing and creative language[8] (see Wanzer et al. 2006) to encourage students by creating an atmosphere of lightheartedness—a sense that learning can be fun or, at least, not utterly serious. And, as prior research has shown, a more relaxed atmosphere can make students more likely to take on difficult tasks and to risk failure (Booth-Butterfield and Booth-Butterfield 1991). The welding teachers' use of spontaneous humor during their one-to-one interactions with students seemed to help produce such an atmosphere.

Playful teasing occurred several times as Stanley ran his welds through the bend-test machine. For a bend test, two 1½-inch strips are cut across a weld, and the two strips are bent in opposite directions so that both the face side and the root side of the weld are tested. Stanley removed the first strip, the face-side strip, from the machine and showed it to Teresa (but first to the camera, see figure 5.15):

Excerpt 5.13

STANLEY: GOOD?

TERESA: [Laughs] I hear angels singing.

STANLEY: [Laughs]

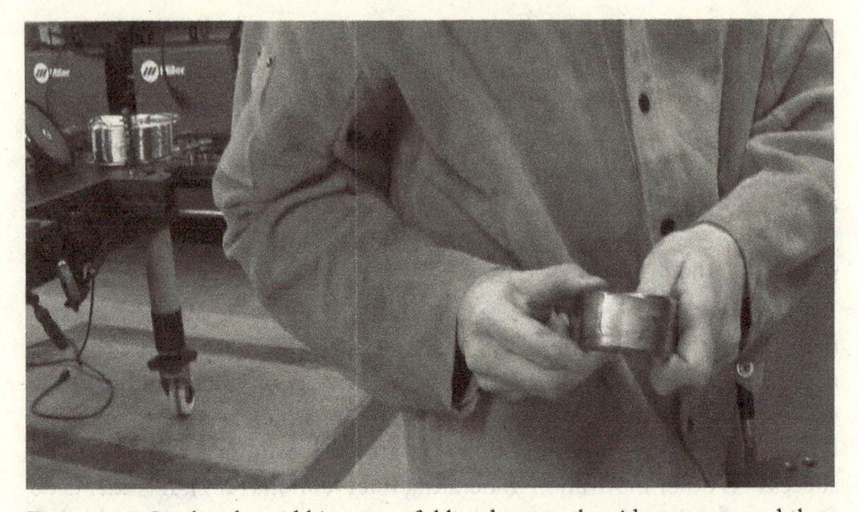

Figure 5.15. Stanley showed his successful bend test to the video camera and then turned to show it to Teresa.

Her response signaled her happiness for Stanley's accomplishment through its teasing dramatization of its significance—even celestial beings had taken note of Stanley's achievement. Teresa's comment calls to mind Bourdieu's (1984) definition of a joke. He wrote that a joke "makes fun" through mockery or insults that are "neutralized by their very excess" (183). He continued, saying that jokes assume familiarly in the freedom with which they exude excess and thus "are in fact tokens . . . of affection" (183). While Bourdieu's definition of a joke may be too limited in scope, it did seem to apply to the excess of Teresa's perception of a choir of angels.

The humor continued as Stanley prepared to bend the root side of his weld, and this time Stanley joined in:

Excerpt 5.13 (continued)
STANLEY: Ok. Roll it through.
TERESA: Ok. No pressure. [Laughs]
STANLEY: A lot of pressure. [Smiling, laughs]

Teresa's comment, "No pressure," played upon the fact that Stanley did indeed feel some pressure to pass this second part of the test; it was the last hurdle to surmount in order to move on to GMAW welding. Indeed, Stanley joked back, "A lot of pressure." But Teresa's humor, generated because it so obviously denied the shared reality of the situation, also helped mitigate its gravity.

Tonya used a slightly different kind of humor at the beginning of her interaction with Simon: feigned jealousy. While the two got themselves into position for Tonya's demonstration of vertical-up SMAW, Tonya observed that Simon's welding helmet (also called a hood) looked like a new and, possibly, expensive one. She joked that she, after years in welding, lacked such a helmet. In response, Simon volunteered that he got it at Emerson, shorthand for Emerson-Fisher Controls in Marshalltown, Iowa:

Excerpt 5.14

TONYA: Hey that's a nice-
SIMON: Yeah?
TONYA: I don't have one like that!
SIMON: I got it at Emerson.
TONYA: Where?
SIMON: Emerson.
TONYA: You got it at Emerson? Yeah, see, I'm going to have to be nice to those guys.

Tonya continued with small talk to suss out why Emerson had graced Simon with such a nice piece of equipment:

Excerpt 5.14 (continued)

TONYA: So you've been at Emerson already?
SIMON: Yep.
TONYA: Did you do an apprenticeship?
SIMON: Yep.
TONYA: For what? Machining? [Mill press?
SIMON: [Machining. We went into weld shopping.
TONYA: Oh really? They- They take you all around huh? Oh that's how you got a hood huh?

Tonya's last question brought the small talk back to the Simon's acquisition of a helmet and renewed her humor. This time, rather than focusing on her feigned jealousy over Simon's possession, Tonya teased Simon, playfully insinuating that his internship at Emerson had been some nefarious plot to obtain a fancy welding helmet. Tonya's use of humor here exemplified what Kotthoff (2009) discussed as "humorous fictions in conversation," fictions created online and, potentially, sustained throughout an interaction (see Demjén 2018). In this case, though, the two got down to business, focusing on how Simon should carry out a tricky corner joint in his weldment.

As my Polish pilots joke illustrates, welding teachers were not the only ones using humor to lighten the mood in the lab. Students did too. And prior research has shown that student humor can generate a variety of benefits, including improvement of students' ability to persevere. In their study of "sojourn" (i.e., study abroad) students' humor, Cheung and Yue (2012) found that students' affiliative humor, humor that makes connections with others by making them laugh, and self-enhancing humor, humor that casts life as humorous (354), mitigated "study hassles," such as meeting supervisors' expectations and writing in English. This mitigation in turn boosted their adjustment to their new location and boosted their resilience (360). Other benefits can arise when students use humor. In a study of college students' ability to cope with job stress, Booth-Butterfield, Booth-Butterfield, and Wanzer (2007) found that students who had scored higher on a test of humor orientation felt that humor could not only help relieve tension but also could help solve the problem they had faced (308). The researchers also pointed out that the very process of generating the humor might focus students outward, toward the recipients of their humor, and away from their own internal distress (308).

A case of student humor occurred as Stein prepared to practice a horizontal open-root joint in SMAW (the same weld that vexed me for about four class periods). Ted had told Stein to tack two coupons together, and then he stepped out of the booth to wait while Stein did as instructed. But after tacking up the coupons, Stein realized that he had gotten into a bind: The bent filler rod that he had used as a spacer between the two coupons was now stuck.[9] After about 45 seconds of trying to pull the rod out, Stein joked about his quandary to Ted, likening his creation to a paperweight with a handle (figure 5.16):

Excerpt 5.15

STEIN: You want a nice paperweight with a good handle on it too?

Stein's self-deprecating humor seemed to be an attempt to relieve the mounting tension that his extended struggle was generating as he fought to get the filler rod out. It also clearly was a call for assistance. And indeed, after another 30 seconds of letting Stein try to solve the problem on his own, Ted moved into the booth beside him to lend a hand.

Figure 5.16. Stein shook his unintended creation, a "nice paperweight with a good handle," up and down in an acknowledgment that he had made a (common) mistake.

Conclusion

This chapter has addressed how teachers attend to the affective component of student learning with motivational scaffolding strategies. These strategies occurred throughout welding interactions, helping to create an environment in which teachers could successfully scaffold students' learning and in which students could actively co-construct their own embodied knowledge. With showing concern strategies, teachers explicitly checked on students' understanding and needs. In doing so, they conveyed that they cared about students and their progress and thus helped build an environment conducive to learning. With humor, teachers lightened the mood. Sometimes, their humor seemed to facilitate learning, as when Tonya used the novel metaphor "beer belly" to describe the drooping weld of Shawn's attempt at vertical-up GMAW (see excerpt 3.6). Her description likely helped Shawn perceive the weld as Tonya did, thus helping to hone his expert perception. Also, perhaps, through its funny, novel metaphor, Tonya's beer belly description may have aided Shawn's recall of holding the arc too long in the weld's middle rather than moving to its edges.

With motivational scaffolding strategies, the welding teachers encouraged students to continue their efforts, especially critical when students were frustrated and struggling. With the giving sympathy strategy, teachers pointed out that the tasks that students perceived as difficult were truly difficult to acquire and acknowledged that the road to competency was a long one for everyone. In this way, teachers alluded to the community of practice that students were entering, pointing out that membership develops incrementally. Whereas giving sympathy acknowledged that the task was difficult, praise underscored students' current successes. Ted's praise for Sawyer's current work in terms of his earlier work was a striking example of this (see excerpt 5.10). Optimism, in turn, predicted students' future successes. Teachers provided evidence for their optimism by referring to past students' progress. They also focused attention on a student's own progress, framing it as prologue to further achievement.

Teachers' gesticulations helped convey their motivational scaffolding strategies. Concrete deictic gesticulations like Tonya's (see figure 5.10) and Ted's (see figure 5.14) identified notable features of students' work—particularly welds—and thus played not only an affective role in scaffolded teaching but also a cognitive one. Iconic gesticulations helped teachers show optimism, as when Teresa illustrated the future welding positions that would be easier for Sophia after she had learned how to perform a horizontal open-root weld (see figures 5.5, 5.6, 5.7, and 5.8). And a metaphoric gesticulation helped Tom convey his sympathy strategy; with it, he illustrated and acknowledged the complexity of welding around a pipe (see figure 5.1). Attending to students' affect, then, teachers relied on nonverbal as well as verbal communication.

In sum, welding teachers' use of motivational scaffolding strategies helped students persevere in their efforts, even through frustration, and thus enabled students' further advancement toward embodied knowledge to take place.

Conclusion

Teaching and Learning Embodied Technical Knowledge

This study examined how, through their verbal and nonverbal technical communication, welding teachers scaffolded students' embodied knowledge and furthered students' membership into what Lave and Wenger (1991) called a community of practice: "a set of relations among persons, activity, and world" (98). For that reason, in this chapter I organize my conclusions around the characteristics of scaffolded teaching: intersubjectivity, ongoing diagnosis, contingency, interactivity, fading, and checking on students' learning. I discuss how teachers used their verbal and nonverbal communication to develop each of these characteristics. I conclude the chapter by discussing what I see as this book's contribution to scaffolding research and then by discussing some avenues for further research on the technical communication of teaching and learning the embodied knowledge of a skilled trade.

Intersubjectivity

The extent to which welding teachers and students' exchanges manifested communication geared toward intersubjectivity—a shared understanding of the task at hand—depended in part on the type of interaction in which they engaged. In the introduction, I outlined four types of interactions that I observed in my study and in my own experience as a student:

1. Interactions that took place when the teacher visited the student's booth on their tour around the lab to check on students' progress.

2. Interactions that took place outside of a student's booth when a student left their booth to approach the teacher with a question or problem or to display a weldment or some other deliverable for the teacher's evaluation.
3. Interactions that took place when the teacher came to the student's booth with a specific pedagogical and preplanned purpose in mind, usually to demonstrate a technique or, sometimes, equipment setup that the student would be attempting for the first time.
4. Interactions that took place when the teacher made a special stop at a student's booth to check on a student because the student had approached the teacher earlier with a question or a problem.

In the first type of interaction, where teachers come to students' booths one-by-one on their rounds around the lab, teachers typically began the process of building intersubjectivity with a question, such as when Tom asked Sebastian, "How we doing Sebastian?" (see excerpt I.1) or with some other prompt. For example, Teresa trailed off to prompt Saul to describe the task he had been working on: "You're working on, um-." The student could then articulate their assessment of their progress, as Saul did ("Went vertical down and then realized I need to go vertical up [unclear] speed") and thus continue focusing the interaction. The teacher could then move to ongoing diagnosis, typically by responding to the student's description of what they had done so far and by evaluating the student's work (typically their welds) in order to gauge the student's current skill level.

In the second type of interaction, ones in which the student approached the teacher outside of their booth, students' questions spurred intersubjectivity, such as Stephanie's question to Tom, "Hey Tom? Tom? Why- what's- what is this?" (see excerpt I.2). This analysis also showed that teachers used the cognitive scaffolding strategy of referring to a previous topic to build intersubjectivity. For example, Tonya's interaction with Sutton started when Sutton approached Tonya outside of Shawn's booth. After Tonya asked "What you got for me Sutton?" Sutton showed Tonya his weld and waited for an assessment of his work. After Tonya confirmed that his weld was acceptable ("I'll buy that one"), she asked whether Sutton had done "up" yet, meaning the vertical-up position. When Sutton answered in the negative, Tonya built a shared understanding of Sutton's next task, vertical up, by referring to a prior lesson about a different weld joint: "You remember how I took your temperatures down when you were going up?" By referring back to a prior lesson on vertical-up welding, Tonya recalled and made relevant a way to set the machine—lowering the voltage for vertical up. Sutton's response in that case confirmed he understood by building on Tonya's reference with a confirmation question: "And move the wire feed down?" (see excerpt 4.9).

In the third type of interaction, intersubjectivity in the lab interaction often stemmed from a prior interaction, often in the classroom, where the teacher would tell the student (or students) that they, the teacher, would demonstrate a new weld joint or position. In such cases, teachers assumed a shared understanding of the task that would be the focus of the interaction and move on to instruction, such as when Teresa described the actions she would carry out in her demonstration, highlighting them for Sophia, and explained their significance, coding them as well (see excerpt I.3):

TERESA: So the whole purpose of this is to be performed between the two plates with a slight drag angle. And what I'm doing is I'm going forward just an eighth of an inch. Coming back and pushing it in. And when I push it in you'll hear a whooshing sound. That whooshing sound's telling me that I'm getting through to the other side.

In such cases, intersubjectivity carried over from a prior interaction and made it possible to move straight into contingent instruction for students who were complete novices in the task, such as, in Sophia's case, horizontal open-root joints. Similarly, in the fourth type of lab interaction, intersubjectivity was already established, based on a prior interaction that determined the focus of the second interaction. In the interactions I observed, teachers and students seemed to have little difficulty in co-constructing an initial shared understanding of what was going on—the process, position, and joint that each student was trying to master.

Of course, once teachers and students had established intersubjectivity, they had to maintain it throughout the interaction, namely, by reinforcing the lesson. For example, Ted reinforced his instruction (a telling strategy), "Get yourself comfortable," later in his interaction with Sawyer with more telling strategies: "But now get yourself comfortable. Hold that tight arm, and follow that along" (see excerpt 5.9). Such examples make clear that intersubjectivity is not a one-time occurrence but an unfolding process. To ensure that their diagnoses are accurate and their interventions are relevant, teachers must make sure that they and their students share an understanding of what is going on.

Ongoing Diagnosis

Except for interactions that took place when the teacher came to the student's booth with a specific purpose in mind (type 3 in the list above), welding teachers diagnosed and monitored students' level of embodied knowledge. Particularly important to this characteristic of scaffolded teaching is teachers' ability to monitor changes in students' understanding. Students do not remain complete newcomers to a task. This study revealed that welding teachers diagnosed students' embodied knowledge via several methods. They asked questions aimed at revealing students' own assessments of their progress. Also, particularly when students came to them with problems, teachers asked questions that asked students to describe what they had done so far. Such questions aimed at sussing out errors in students' setup or (self-described) technique.

Teachers also observed students to assess whether they were using proper technique. The importance of such observations came up in one of my interviews with Tom. He pointed out that observations of students as they weld allow him to home in on problems with students' technique, whereas ambiguity remains after analyses of students' completed welds:

TOM: The big one students tell me that they don't like is they'll show me a weld and say, "This is bad. How do I get better?" And looking at the weld, there can be multiple causes for the result that they got. So then I'll just tell them, "Weld for me. Let me watch what you are doing."

When teachers observed students as they welded, they typically toggled between diagnosis and the "calibrated support" (Puntambekar and Hübscher 2005, 2) of contingent intervention—typically telling strategies, such as Tom's instruction to Sebastian (see excerpt 1.1):

TOM: [Sebastian welds, 10s] That speed looks good. [2s] Get just a little bit closer to the plate. [6s] Back up a little bit. Stop.

But teachers also provided intervention after students had completed a weld, as Tonya did (see excerpt 3.13), as Ted did (see excerpt 4.7), and as Tom did too after observing Sebastian:

TOM: I think you might have sped up. Just a touch.
SEBASTIAN: Ok.

It's not clear whether the timing of feedback—whether teachers intervene during the task or immediately after—affects student learning. Giving feedback during a weld has the advantage of immediate support from moment-to-moment diagnosis. Giving feedback after a weld (or the performance of some other task) has the advantage of allowing the student to concentrate on the feedback as opposed to focusing on performing the task.

As Tom alluded to in his interview, ongoing diagnosis also occurred through teachers' evaluation of students' deliverables—mainly their weldments but also their other work, such as bend tests. In interactions that occurred during teachers' circuits around the lab from booth to booth and in interactions that occurred outside students' booths, teachers commonly used their expert vision to evaluate students' work and use their assessments as a foundation for a contingent intervention, often instruction but also pumping questions and praise. In support of their assessments of students' work, teachers used concrete deictic gesticulations in particular. For example, as she praised Steve's weld, Tonya used a concrete deictic gesticulation to point precisely to the section of Steve's weld that was, in her estimation, both "perfect" and "beautiful" (see figure 5.10). That is, they used concrete deictic gesticulations to highlight the phenomenon that the student should notice (see also, figure 4.7).

Contingency

With a contingent intervention in a student's learning, the teacher adjusts knowledge so that the knowledge engages the student at their current level of understanding, or in their ZPD. This study has examined the strategies that welding teachers used to intervene in students' learning once they had diagnosed students' level of understanding. One of the most important of these was demonstration, the nonverbal instruction strategy with which teachers showed students how to perform some task. The scope of the task varied. A teacher could demonstrate a complete weld, typically by running a weld to join two coupons across their length. Sometimes teachers demonstrated just a part of a task, showing the correct travel and work angle or showing an effective body position. Sometimes teachers completed just part of a weld run, leaving the rest for the student to complete. Stein, for example, described a lesson in which Ted alternated between demonstration and observation:

STEIN: Like, Ted on Monday came in my booth. He ran half a bead and then he showed me basically what to do, and then I did right after him. . . . I'm more hands on, so like, an instructor shows me something and then they watch me do it, I like that a lot better. . . . because Ted's been doing this for a while- for a long time now, and he knows his stuff. And if he tells you to slow down, slow down, speed up, speed up. It works out perfect. I like that.

Besides demonstrating welds, teachers also demonstrated how to use and maintain equipment, such as plasma cutters, and tools, such as grinders. Such interventions occurred when teachers determined that students needed instruction; that is, when students needed direct and substantive support to proceed in developing their embodied knowledge.

The timing (i.e., the temporal contingency) of teachers' demonstrations varied according to students' familiarity (or lack of it) with the task and according to other demands placed on the teachers' time. For example, teachers used demonstrations when students were complete novices to a task, sometimes gathering more than one student together to observe the demonstration. Students watched as the teacher performed the task and then right after attempted the same task themselves (for examples, see excerpts 3.12 and 4.4). In such cases, teachers engaged in what McLain (2018) called "just-in-time" demonstration. McLain contrasted just-in-time demonstration against "frontloaded" demonstrations, which involved a delay between the teacher's demonstration and the student's attempt at the task. For example, welding students might watch a video demonstrating the steps of a particular weld joint but end up not attempting the task until the next class period. From my observations and experience, however, welding teachers relied far more heavily on just-in-time demonstrations, which have the advantage of minimizing students' need "to encode complex information to long-term memory" and thus lowering their cognitive load (McLain 2018, 987). The efficacy of frontloaded demonstrations, in contrast, rests on students' ability "to translate observations from working to long-term memory" (987). Thus, the timing of welding teachers' demonstrations seemed well placed.

Besides using the instruction strategy of demonstration when students were newcomers to a task, teachers also employed demonstrations after students had already attempted—and struggled or failed—at its correct completion. McLain (2018) called such interventions "after-failure" demonstrations. After-failure demonstrations, according to McLain, make way "for exploration and trial and error" (987). It seemed that the timing of after-failure demonstrations was particularly critical, as students' frustration level could grow and inhibit motivation to persist in the task. After I had failed over and over to complete a horizontal open-root weld joint, my SMAW teacher, stopping on her circuit of the lab, demonstrated the weld for me. After even more failures to complete even one entire run without blowing a hole in the joint by traveling too slow or without complete penetration by traveling too fast, my GMAW teacher demonstrated the task yet again. He had been walking through our lab on the way to his lab down the hall and, luckily for me, had stopped to say hi. After these demonstrations and after hours of practice, I finally obtained one satisfactory horizontal open-root joint.

This study revealed two types of welding demonstrations: ones that actually involved striking an arc and welding, and ones that did not. In the latter, teachers demonstrated an action by mimicking it. When teachers modeled correct body positions and techniques without running a weld bead, they were better able to speak normally, as opposed to speaking from under a welding helmet and over the sound of laying a weld. However, overlaying the former type of demonstration with verbal strategies was not impossible (see excerpt 4.6)—just more challenging. To complement the visual instructional intervention of demonstration, then, teachers could provide verbal instruction. In doing this, they provided a multimodal understanding of the task being learned.

This study revealed that during demonstrations, teachers often paired describing and explaining strategies. Teachers' describing strategies highlighted, or identified, phenomena, such as a groove near the toe of the weld, and their explaining strategies coded, or interpreted, the significance of those phenomena, such as undercut caused by too-fast travel speed. These describing-explaining pairs, then, combined with teachers' demonstrations, seemed rich—though directive—sources of support for students' progress toward embodied knowledge. Indeed, Sam noted that teacher demonstration combined with describing and explaining strategies helped him understand how to "read" the puddle:

SAM: Oftentimes they'll [teachers will] come into the booth and just- I'm struggling with something- show me how it's done. And I'll watch the puddle as they're manipulating it and be able to see, ok, that's what it's supposed to look like. . . . Oftentimes they'll come in and start welding, and I'll be watching and they'll be talking and telling me what's going on with the puddle. Describing how it's working. Especially with one time with stick. We were doing vertical up. And my instructor was telling me about watching the slag at the top of the puddle. It's indicating something. And so it's just the explaining while I was seeing it is really helpful for me.

These comments suggest that students appreciated teachers' ability to combine demonstration with verbal strategies, a competency that, as Tom noted, can take some time to get one's brain "wired to the point" where it's possible to weld and talk about welding simultaneously. Future research might explore methods to facilitate training in this skill—one that is incredibly important for teachers of embodied knowledge, such as welding.

Teachers' describing strategies (e.g., "So it's starting to flatten out") were noteworthy for another reason: Within them, teachers sometimes employed metaphors. In my informational interviews with the four welding instructors in this study, I asked them about whether they deliberately employed metaphor in their teaching (see Deignan 2011; Thonus and Hewitt 2016, 53). They all thought that they did, but only Teresa was able to bring to mind a specific example of how she has used a metaphor to describe a particular welding technique:

TERESA: I do- If they're doing a Z weave with a seventy eighteen [a 7018 electrode] or something, it tends to get real hot on the last inch and a half. And you really have to increase your travel speed. You can't go up farther. And this one doesn't work anymore. It only works with older students. I've noticed I talk about the VCR. . . . You have to hit the fast forward on the VCR. . . . With the younger students I usually use the analogy that you're running a race and you have to sprint right before the finish line.

Teresa's comment suggests that she had reflected upon the source domain that she employed, considering whether the students showing up in her lab had ever used a VCR—or even seen one in person.

In contrast, Ted's recollections went to a metaphor that he used to discuss the difference between SMAW and GTAW:

TED: Yeah. I do use stuff like that. I mean, I always say stick is like driving your F150 truck, and TIG is like driving your Ferrari.

While the welding teachers couldn't think of many metaphors that they employed in their classes, this study's discourse analysis revealed that they nevertheless employed a variety of novel metaphors, what Goschler (2019) called "creative metaphors" (82), to scaffold students' learning. Goschler said that such novel metaphors "are neither typical for everyday language nor for the terminology of the respective discipline" (82). That is to say, a metaphor like Tonya's "beer belly" differs from a conventionalized, disciplinary metaphor like the "toe" of a weld. Such novel metaphors, whether deliberate or not, "encourage the students to link the novel concept to their embodied experience" (80). Goschler studied the metaphors of history and chemistry teachers, but much room exists for in-depth research of metaphor in the technical communication of skilled-trade teaching. Such research seems particularly important given that, in the field of technical communication, research on metaphors has focused mainly on their use to communicate science (e.g., Baake 2003; Giles 2008), rather than to teach embodied knowledge.

This study also showed that teachers used what I have called negative demonstration, showing students what *not* to do. This finding confirmed what Asplund and Kilbrink (2018) articulated in their study of the talk between a welding teacher and student in a Swedish upper-secondary technical school. In that study, the welding teacher combined a telling strategy and a demonstration to show a gun angle that would not be effective and how that teacher contrasted ineffective technique against effective technique. The present study, too, showed that welding teachers drew upon their experience and its concomitant expertise to isolate body positions and techniques that newcomers commonly get wrong, anticipating a student's mistakes in the hope of helping that student to avoid them. In addition, this study showed that welding teachers used negative demonstrations in after-failure demonstrations after watching students perform a task. Their negative demonstrations served as a sort of instant replay with the goal of making salient students' mistakes, such as Simon's travel angle (see excerpt 3.13).

Of course, welding teachers didn't demonstrate during every interaction with students. Even so, visual communication remained an important part of how they scaffolded students' learning. Specifically, when teachers intervened with verbal strategies, they used gesticulations as well to help convey their meaning. Teachers frequently used concrete deictic gesticulations with instruction strategies, such as telling strategies (see figure 4.1, for example) and with pumping questions (see figure 4.7, for example). With pumping questions, teachers provided less support than they did with instruction and thus left more room for students to assume responsibility for their own learning. This study revealed that teachers used pumping questions in particular when they scaffolded students' expert perception, that is, when the goal was to highlight a phenomenon for students, thus simultaneously helping to construct intersubjectivity. Concrete deictic gesticulations worked in tandem with teachers' pumping questions, as when Tom pointed to the precise location on Stephanie's wrist that he expected her to feel discomfort (see figure 4.15), thus supporting the development of her ability to perceive like an expert welder. Concrete deictic gesticulations helped make teachers' intended referents extremely clear and allowed them to differentiate between the object or location that they intended and those they did not. Teachers often paired pumping questions with explanations, using the less directive strategy to create intersubjectivity and then the more directive explaining strategy to help students understand the phenomenon.

This study showed that concrete deictic gesticulations also played a role in teachers' contingent motivational scaffolding. Assessing through ongoing diagnosis not just students' level of understanding but also their affective state, particularly their level of frustration, teachers employed strategies such as praise to maintain students' motivation. Specifically, with concrete deictic gesticulations, they identified the referent of their praise. For example, Tonya pointed to the section of Steve's weld that was "beautiful" (see figure 5.10), highlighting the best result that Steve had achieved on the weld run. Ted pointed at Sawyer's weldment to identify Sawyer's progress: "Yeah, now look at them runs you got in there. You had three four really good runs" (see figure 5.14). In addition, this study also showed how teachers withdraw their concrete deictic gesticulations as they moved from negative evaluation to motivating praise. Teresa pulled in her pointing gesticulation as she switched from assessing Susan's weld with "So you went a little slow, right here" to providing a positive assessment overall with "but most of it looks good" (see figure 5.11). Teachers' concrete deictic gesticulations helped them highlight phenomena, and when they accompanied praise, supported students' affect as well.

In contrast, with abstract deictic gesticulations, teachers could refer to nonpresent objects or abstractions, including referring to future and thus presently nonpresent objects, such as a student's future weldment. This finding coincides with Haas and Witte's (2001) analysis of the gestures that city employees used as they drafted technical standards for city development. They found that city employees used concrete deictic gesticulations, what they called "literal pointing," when they referenced an entire or a part of a technical drawing and abstract deictic gesticulations, what they called "nonliteral gestures," when they referred to objects in the world that the drawing represented (443). The city employees' abstract deictic gesticulations also referred to objects present in the world but not present in the meeting room, such as a channel or an overhang (443). Teachers (and students such as Seth too, see figure 3.10) used abstract deictic gesticulations to refer to future outcomes.

This study showed, for example, how teachers used metaphoric gesticulations to represent characteristics of abstractions such as speed, fluidity, size, and width. For example, talking to Stephanie, Tom used a metaphoric gesticulation, representing arc width (see figure 5.3). In contrast, teachers used iconic gesticulations to represent the objects or actions themselves. For example, talking to Sophia, Teresa used iconic gesticulations to represent vertical-up and overhead welds (see figures 5.5, 5.6, 5.7, and 5.8). In short, at times, teachers alternated between iconic and metaphoric gesticulations to move from representing specific objects to abstracted characteristics of those objects in a variety of verbal strategies, but particularly the instruction strategy of explaining.

Finally, in relation to teachers' contingency, specifically their explaining strategies, I wondered through my own welding experience and through my observations and analysis whether welding teachers might more often have explained the common uses of various weld joints and positions. On a few occasions, my teachers told me about odd situations that they had encountered on the job. For example, my FCAW teacher told me about a time he had to overhead weld while lying on his back. That story revealed the importance of learning how to weld overhead. But I would have benefited, I think, from more explanations, particularly explanations of common—as opposed to atypical—applications of the weld joints and positions that we learned.

Interactivity

A fourth element of the scaffolding process that Puntambekar and Hübscher (2005) outlined is interactivity, a primary means through which teachers engage in ongoing diagnosis and provide feedback through which students ask questions, respond to teachers' prompts, and raise new topics. About interactivity, they wrote:

> Although the teacher plays a vital role in the instructional process, the [student] is also an active participant so that scaffolded interactions are a function of participation by the teacher and the learner. The dialogic interactions . . . enable the teacher to conduct an ongoing assessment of the student's understanding and allow the student to play a role in negotiating the interactions. (3)

The idea behind interactivity is that it is ongoing and reciprocal. But as the expert, the old timer in Lave and Wenger's (1991) terms, teachers carry the responsibility of facilitating students' participation.

The pumping questions that the welding teachers used to develop students' expert perception are perfect examples of interactivity. Through pumping questions such as Teresa's "See how uneven these plates are?" (excerpt 4.3), teachers highlighted cues in the environment that would allow students, in the future, to identify for themselves critical sensory signals, including ones that indicated their level of success in the task. For example, Tonya asked Steve "See how the puddle's getting ahead of you?" (excerpt 4.2). After making phenomenon salient, teachers could move to coding, generally via an explaining strategy.

Besides scaffolding students' learning, teachers' pumping questions also pushed students to participate in their interactions, to actively engage in constructing their own meanings, for example, the cause-and-effect relationships between their actions and their results. However, teachers' pumping questions ranged in the extent to which they constrained students' responses. Most required just a "yes" or "no" response from students. Even so, even close-ended pumping questions such as yes-no pumping questions provided an opening—one that students could potentially fill with a more-than-minimal response. That said, students would likely have benefited from broader, more open-ended pumping questions. Such questions pushed students to more substantive responses.

Finally, in addition to asking more open-ended questions, teachers might use their questions to ask students to code, to explain, some phenomenon in the environment. Teresa did just that when she asked Saul to evaluate the penetration on his weld: "How was the penetration?" (see excerpt 4.3). As Dix (2016) wrote, "Worthwhile questions are regarded as those that challenge students' thinking, deepen understandings and promote reflection, analysis, self-examination and inquiry" (24). Thus, teachers who attend to the types of questions that they ask can better match their question to their purpose, for example, determining whether a student perceives a phenomenon or prompting a student to assess their results.

This study showed that students' questions generated interactivity as well. Perhaps most important were the questions they asked to fill in gaps in their knowledge, such as Stephanie's question to Tom, "Hey Tom? Tom? Why- what's- what is this?" As excerpt I.2 showed, Stephanie's question set into motion an extended explanation and advice for how she should proceed. Like Stephanie approached Tom outside of her booth to ask a question about powdery buildup on her weld, Susan approached Teresa to get advice for improving her results, at one point asking, "Aim at the toe or I will aim a little above the toe?" With their questions, students helped teachers provide a contingent intervention, often instruction. Research in education has examined students' questions (e.g., Park 2017; Skarbø Solem 2016) and has suggested that students may avoid questions for fear of being impolite (e.g., for fear of insinuating that the teacher has not adequately done their job). In addition, as Person et al. (1994) delineated in their study of student questions, students incur costs when they ask questions, including "revealing ignorance and losing status" with a "bad question" and "not being viewed as an independent problem solver" (207). The welding students that I observed seemed willing to risk asking questions to get the support that they needed, likely because their teachers had made them feel secure in doing so. Indeed, as van de Pol et al. (2012) discussed in their study of teachers' scaffolding in ninth-grade social studies classes, students considered "an open and encouraging tone . . . to be important in creating a safe atmosphere" (199). The one-to-one, private nature of students' conversations with their teachers likely facilitated welding students' questions. Indeed, in his postinteraction interview, Sam told me, "I really like the one-on-one time with the instructor. Where it's just undivided attention and I can ask funny questions or ridiculous questions that I want to understand. They have the time to explain."

Even more commonly, it seemed, welding students asked confirmation questions, questions that students posed to check their understanding of something the teacher had said. Multiple excerpts throughout this analysis have included student confirmation questions, such as Seth's question to Teresa, "With a higher wire speed?" (see excerpt 3.2) and Shawn's question to Tonya, "Did you say beer belly?" (see excerpt 3.6). These questions, too, generated the interactivity that in part characterizes scaffolded learning, as they signal students willingness to climb the scaffold under (co) construction.

Finally, before leaving the topic of interactivity, it is important to note once again the interactivity generated as teachers and students engaged in joking and other humorous behaviors. In chapter 6, I discussed how an exchange of jokes that I had with Ryan and my GMAW teacher made me feel like an in-group member, part of what Fine (1987) called an idioculture (125). Friendly exchanges such as that one did not necessarily pertain to welding and the development of embodied knowledge, but they could, as excerpts 5.13 and 5.15, for example, demonstrate. But they did help generate the open atmosphere that van de Pol et al. (2012) discussed and facilitated open exchanges and thus learning. That said, as prior research has quite clearly demonstrated (e.g., Ford et al. 2014), humor can also "reinforce social, ethnic, and cultural identities and inequalities" (Van Praag, Stevens, and Van Houtte 2017, 393). For example, I described in chapter 2 how one of my teachers mocked my note-taking and my height and in the process made me feel like an interloper in a classroom filled with young men. The humor that I observed from the teachers at NTC, LSC, and MCC, however, seemed intended and perceived as well-meaning and friendly banter.

Fading

As Puntambekar and Hübscher (2005) discussed, fading refers to the teacher's incrementally giving over to the student the responsibility for the task. In chapter 2, I noted how Ted's "crawl, walk, run" method for learning machine setup constituted a sort of fading. In the first week, Ted said, he demonstrated how to set up the machine. In the second week, he demonstrated it again, but he required students to write down a list of the steps. In the third week, students set the machines up themselves, using the instructions that they had written for themselves. Ted thus faded at a macrolevel, giving over responsibility for a task over a span of time substantially longer than the duration of a single interaction. But this study showed that fading occurred at the microlevel as well, within the span of a single one-to-one interaction between a welding teacher and student. However, as mentioned previously, fading in welding teachers' interactions meant ending the interaction by providing advice for the student's next attempt, usually unsupervised as teachers needed to circulate around the lab so that they were available to all students.

For example, in relation to excerpt 5.9, I discussed how Tonya ended her interaction with Shawn by combining praise ("But that was really good") with a critique of the work she had seen thus far ("you're just a little too much in the middle"). Later, after this praise-criticism pair, Tonya parted ways with Shawn by imparting advice for his next try: "Hold on those sides. It will flatten out for you." Tonya withdrew from her interaction to allow Shawn time to practice on his own, but before she did, she intervened with instruction, telling strategies, directing him in how to improve his results.

Fading, then, assumed that students would continue to practice on their own, as telling strategies such as Teresa's to Seth, "Run a couple and see how it goes" (see excerpt 3.11). Such parting directives indicated teachers' assumption that they would check on the student later and engage once again in an iterative cycle of intersubjectivity, contingency, interactivity, and fading. In the case of welding teachers and students in this study, fading occurred over multiple interactions, as students developed their ability to complete a task successfully on their own.

Checking Students' Learning

By checking on students' learning, van de Pol et al. (2012) wrote, "teachers attain feedback on the effectiveness of their scaffolding efforts" (203). However, they found that time constraints kept the teachers in their study from checking back with students. Time constraints govern the welding lab too. Welding teachers could check on students' progress if they had time for a second tour around the lab. The time they had depended on myriad factors, including students' need for time-consuming demonstrations, the need to help students change out gas tanks or wire spools, and the number of times students approached them with questions and requests for weld inspections as they made their tour around the lab.

Frequently, then, checks of students' progress occurred when students left their booths and found their teacher elsewhere in the lab (type 3 in the list at the outset of this chapter). In other words, checking occurred, but in the welding lab it often relied on a student's initiative to seek out their teacher for further diagnosis and contingent intervention.

Contribution to Scaffolding Research

The study I have reported in this book follows the work of others scholars who have for a long time now tried to isolate and describe the strategies that teachers use to scaffold students' learning. Such researchers have tried to answer the questions: what actually occurs in teachers' and students' talk during scaffolded teaching and learning? For example, studying scaffolding in literacy teaching, Pinnell et al. (1994), pointed out that tutors prompted students and offered praise (22–23). Graesser et al. (1995) and Merrill et al. (1995) examined tutors' questions, including pumping questions like the kind found in the welding teachers' talk. Chi et al. (2001) examined the role of tutors' prompts and explanations in tutoring about the human circulatory system. More recently, Mackiewicz and Thompson (2018) built on Cromley and Azevedo's (2005) work to develop a scheme of strategies that helps explain how writing tutors enact scaffolded teaching. As I discussed in chapter 2, for the present study I modified Mackiewicz and Thompson's (2018) scheme to analyze scaffolding during pedagogical interactions about welding.

Perhaps the most important modification to that scheme for use in different—a more embodied—context, was the addition of a describing strategy—one of five instruction strategies. In using this strategy, teachers related the characteristics of a thing or action, occasionally with a metaphor. Prior analyses of scaffolded teaching and learning have overlooked this critical means by which teachers scaffold students' ability to see, hear, and touch/feel as experts. With describing strategies—often used in conjunction with demonstrating strategies—teachers highlighted important phenomena, setting them apart from other, less important phenomena in the shared environment. For example, in excerpt 4.1, Tom described his travel speed as he demonstrated: "You can see how slow I'm moving." Teachers' describing strategies often laid a groundwork for their explaining strategies—the means by which teachers imbued those phenomena with meaning, as when Tom explained the meaning of the arc's size: "So I'm going to try and shrink that down so we have a little more control in the groove."

In sum, then, this study contributed to research on scaffolding, particularly in pedagogical contexts where teachers and students co-constructed embodied knowledge, in that it pointed out the importance of describing critical phenomena in the environment to help students grow in their ability to perceive as welders.

Possibilities for Future Research

I hope that this research project on the verbal and nonverbal communication that welding teachers used in their scaffolded teaching constitutes the beginning of further research to better understand the technical communication that drives the development of embodied technical knowledge, particularly in the skilled trades. This study focused on welding, but further research might focus on the communication surrounding the learning of other embodied fields, such as carpentry, tool and die manufacture, automotive and motorcycle mechanics, and electrical work, to examine commonalities across teachers' scaffolding strategies and students' entry into community-of-practice membership. Some such research has already begun, such as de Saint-Georges and Filliettaz's (2008) study of automotive teaching and Asplund and Kilbrink's (2018) case study of a plumbing student learning how to weld. Such future research might investigate the relationship between the teachers' and students' classroom interactions and their interactions in the lab. For example, when teachers use the cognitive scaffolding strategy of referring back to a prior lesson, they will likely reference lessons they imparted during lectures or demonstrations that students watched on videos. A future study might more carefully attend to the technical communication being co-constructed during skilled-trade teachers' and students' interactions outside of the lab.

Future research also might more carefully examine the role of racial and gender identity on teaching and learning a skilled trade and the verbal and nonverbal communication that construct that teaching and learning. This study reported some of the experiences with sexism reported by the women who participated in this study. No doubt, though, future research taking a more robust approach to gender identity—and racial identity as well—could do better justice to the complexity of navigating entry into and a subsequent career within a skilled trade such as welding.

In relation to teachers' use of specific teaching strategies, future research might further investigate teachers' use of novel metaphors, such as Teresa's "dry rivers" and Tonya's "beer belly" metaphors. Such research should also investigate the effect of novel metaphors on students' understanding and recall of a given concept. It might also investigate the extent to which teachers and students continue to use teachers'—and perhaps students'—novel metaphors. Also, future research might focus on the timing of teachers' demonstrations, examining the outcomes of what McLain (2018) called just-in-time and after-failure demonstrations.

The next step in my research is to begin a study of the potential role of multimodal composition in skilled-trade education. As mentioned in the introduction, Ted mentioned that he asked students to list the steps of setting up a welding machine in order to learn how to carry out this procedure on their own. Tom also touched on the utility of writing. He said that he incorporated writing into his welding classes, for example, in students' assessments of their welds, because he believed that writing would help them evaluate their own work and solve problems that they identified:

Tom: And then if they saw issues with the weld, so if they had discontinuities or other things like that, they were documenting those as well. And then they were writing down how they were going to fix it. And it was one of the ways for me to kind of have- document what they were doing so that I could track it, and then just to really force them to actually think through it that way.

I intend to begin with an ethnographic study, examining the types of texts that skilled tradespeople use on the job. My end goals are these: to rethink writing courses for students in skilled-trades programs so that those courses better engage students and further their career goals and to help welding teachers generate writing assignments that facilitate students' understanding of important concepts and skills.

In relation to my life as a welding student, well, I began as a complete novice, a student struggling with her body's inability to use a clamp, among hundreds of other minor humiliations and frustrations. I end this book as someone only marginally more adept with clamps. But I also leave it as someone who knows, for example, how to lay a fillet weld in a T-joint, how to use a grinder and a plasma cutter, and how to change out a wire spool on a welding machine. That's to say, I end this book as someone continuing to grow in her technical knowledge, embodied.

Appendix A

Informative Interview Questions for Teachers

Demographic Information

- Years of experience in on-the-job welding: ________________
- Years of experience in teaching welding: ________________
- Years of experience in teaching welding at this school: ________________
- Native language: ________________
- Native country: ________________
- Age: ________________
- Gender: ________________

Potential Questions

1. What do you find most satisfying about teaching welding?
2. What courses do you prefer to teach and why?
3. If you could change one thing about the welding curriculum here, what would you change?
4. When you're teaching, do you think about how you are describing correct technique?
5. What sorts of strategies do you use to show students correct techniques?

6. Do you use analogies or similes when you are teaching? If so, how?
7. Do you demonstrate correct techniques for students? If so, how often?
8. To what extent do you incorporate the use of blueprints and welding symbols in your courses?
9. How do you organize a course? What should go at the beginning? What should go at the end?
10. More women are entering welding programs. Has this shift changed how you communicate in the classroom/lab?

Appendix B

Postinteraction Interview Questions for Teachers and Students

Teacher Demographic Information

A. Years of experience in on-the-job welding: ________________

B. Years of experience in teaching welding: ________________

C. Years of experience in teaching welding at this school: ________________

D. Native language: ________________

E. Native country: ________________

F. Age: ________________

G. Gender: ________________

Potential Questions for Teachers

1. What do you find most satisfying about teaching welding?
2. To what extent do you think about how you teach welding?
3. When you're teaching, do you think about how you are describing correct technique?

4. What sorts of strategies do you use to show students correct techniques (in the classroom and in the lab)?
5. Do you use analogies or similes when you are teaching? If so, how?
6. Do you demonstrate correct techniques for students? If so, when?
7. What concepts or techniques do students struggle with? How do you try to help?
8. More women are entering welding programs. To what extent has this shift changed how you communicate in the classroom/lab?
9. In this particular interaction, what were you trying to accomplish with the student?
10. To what extent is what you said and did in this interaction typical? In what ways does it differ from what you usually do?
11. To what extent do you think this interaction was successful? Do you think the student understood what you were trying to convey? How can you tell?

Student Demographic Information

Years of experience in on-the-job welding: ________________

Months or years of experience in studying welding: ________________

Years of experience in studying welding at this school: ________________

Native language: ________________

Native country: ________________

Age: ________________

Gender: ________________

Potential Questions for Students

1. What do you find most satisfying about studying welding?
2. To what extent do you think about how you learn welding? What methods do teachers use that seem to work best for you?

3. To what extent do you find it helpful when your instructor describes correct technique?

4. To what extent do you find it helpful when your instructor demonstrates correct technique?

5. What concepts or techniques have you struggled with? How have instructors tried to help?

6. More women are entering welding programs. To what extent has this shift changed how you communicate with other students in the classroom/lab?

7. In this particular interaction, what were you trying to accomplish (i.e., what were you learning)?

8. How did your instructor try to help you learn this process or position?

9. To what extent do you feel that you learned this process or position?

10. To what extent is what you and your instructor did and said in this interaction typical? In what ways does it differ from what instructors usually do?

11. To what extent was this interaction successful? Did you learn what you wanted to learn?

Appendix C

Informed Consent Form for Welding Teachers and Welding Students

Title of Study: Spoken and Signed Communication in Welding Interactions
Investigator: Dr. Jo Mackiewicz, Iowa State University

This form describes a research project. It has information to help you decide whether or not you wish to participate. Research studies include only people who choose to take part—your participation is completely voluntary. Please discuss any questions you have about the study or about this form with the researcher, Dr. Jo Mackiewicz, before deciding to participate.

Introduction

The purpose of this study is to examine the communication that surrounds learning and teaching how to weld. You are being invited to participate in this study because you are an instructor or a student in a welding program. You should not participate if you are under 18.

Description of Procedures

If you agree to participate, I will video record you as you teach or learn how to weld. I will also ask you to participate in an individualized, one-to-one interview (about 20 minutes) to discuss the video-recorded interaction. I will audio record the interview.

Risks or Discomforts

While participating in this study you may at first feel uncomfortable about being video recorded (in the welding lab) and audio recorded (in the following interview) because you may feel like you are being watched or judged. Please note that my intent is not to judge your learning or teaching, and I will not be watching you through the video recorder. Also, it is often the case that this discomfort wears off as you concentrate on the task at hand.

Benefits

It is hoped that the information gained in this study will benefit society by revealing aspects of communication that make it easier to teach and learn how to weld.

Costs

You will not have any costs from participating in this study.

Participant Rights

Participating in this study is completely voluntary. You may choose not to take part in the study or to stop participating at any time, for any reason, without penalty or negative consequences. Most important for students to note is that participating in this study or choosing not to participate in this study will not affect your grade in this class or your standing with your instructor.

If you have any questions *about the rights of research subjects or research-related injury*, please contact: The IRB Administrator, (515) 294-4566, IRB@iastate.edu, or Director, (515) 294-3115, Office for Responsible Research, Iowa State University, Ames, Iowa 50011.

Confidentiality

To ensure confidentiality to the extent permitted by law, the following measures will be taken:

- All consent forms will be scanned into a secure CyBox account (a secure file-storage system administered by Iowa State University) and then shredded.

- All video recordings, audio recordings, and transcripts will be kept in a secure CyBox account (a secure file-storage system administered by Iowa State University).

- No names will be attached to video recordings, audio recordings, or transcripts of recordings. Instead of participants' names, a coding system for participants will be used (e.g., Tom and Sebastian). This process is part of de-identification.
- In any video clips or photographs appearing in publications resulting from this research, faces will be blurred out so that unique individuals are not identifiable. This process is part of de-identification as well.

- Please note that I may share with other researchers de-identified information that I collect about you during this study. I may also use the information in future research studies. I will not obtain additional informed consent from you before sharing the de-identified data.

However, federal government regulatory agencies auditing departments of Iowa State University, and the Institutional Review Board (a committee that reviews and approves human subject research studies), may inspect and/or copy study records for quality assurance and data analysis. These records (i.e., scanned consent forms) may contain private information.

In addition, I cannot completely guarantee confidentiality due to the relatively small number of people involved in the welding programs from which I am gathering data. However, to help reduce this concern, in any publications resulting from this research, I will not link individual code names to individual schools.

Sharing Results

I will share reports of this research with you upon request. You can contact me at jomack@iastate.edu or xxx-xxx-xxxx.

Questions

You are encouraged to ask questions at any time during this study. For further information *about the study,* contact Dr. Jo Mackiewicz (jomack@iastate.edu).

Consent and Authorization Provisions

Your signature indicates that you voluntarily agree to participate in this study, that the study has been explained to you, that you have been given the time to read the document, and that your questions have been satisfactorily answered. You will receive a copy of the written informed consent prior to your participation in the study.

Participant's Name (printed) ______________________________

______________________________________ _______________

Participant's Signature Date

Appendix D

Solicitation Message for Students

Dear ______________________ :

I am writing to ask you to consider participating in my research study about the communication that surrounds teaching and learning how to weld.

I'm a professor in the English Department at Iowa State University. I am also a student in the welding program at DMACC (in Ankeny, Iowa).

To study the communication that surrounds welding, I am asking to video record you as you learn how to weld in the welding lab. Specifically, I am asking to video record you as your instructor teaches you a new process or position, which should take about 10 to 15 minutes. I am also asking to audio record an interview with you about the video-recorded interaction. The interview would take about 20 minutes, and I would work around your busy schedule.

I would not use your name in any publications stemming from this research project.

Thank you for considering my request. I'd be happy to answer any questions that you have. My email and phone number are in my signature below.

Sincerely,
Dr. Jo Mackiewicz

Glossary

AC A current of electricity in which electrons regularly change direction.

amperage A unit measuring the rate that electrons flow through a conductor, such as an electrode.

bend test A method of testing a weld's ductility and resistance to fracture. The test involves cutting two 1½-inch cross-sections from a weld. One section is used to test the face side of the weld, and the other is used to test the root side. Each section is bent in a jig.

DC– A current of electricity in which electrons flow in one direction. In welding, the electrode connects to the negative terminal of the power source, and the work connects to the positive terminal.

DC+ A current of electricity in which electrons flow in one direction. In welding, the electrode connects to the positive terminal of the power source, and the work connects to the negative terminal.

ductility The quality of being able to stretch under tensile strain.

ESAB A brand of welding equipment. The acronym is short for Elektriska Svetsningsaktiebolaget, which means Electric Welding Limited company. ESAB produces welding machines that are yellow.

FCAW An abbreviation for flux-core arc welding. FCAW differs from GMAW in that the gas shielding the arc comes from a flux inside the filler wire.

fillet weld A weld that joins two perpendicular pieces of metal. Fillet welds are used for T-joints, lap joints, and corner joints.

flux A substance (e.g., carbonate and silicate materials) used in welding processes, namely, FCAW and SMAW, to shield the weld from atmospheric gases. Flux melts and produces gas that pushes atmospheric gas back and thus prevents oxidation of the weld pool.

GMAW An abbreviation for gas-metal arc welding. GMAW is also known as MIG welding. GMAW passes metal filler wire from a spool through the welding gun, or torch. An electric arc, shielded by a gas mixture (e.g., 75% argon and 25% CO_2), forms between the filler wire and the base metal.

GTAW An abbreviation for gas-tungsten arc welding. GTAW is also known as TIG welding. GTAW uses a tungsten electrode to create an electric arc.

lap joint A joint between two overlapping pieces of metal.

Lincoln A brand of welding equipment. Lincoln welding machines are red.

MIG welding An abbreviation for metal-inert gas welding. MIG welding is also known as GMAW. MIG welding passes metal filler wire from a spool through the welding gun, or torch. An electric arc, shielded by a gas mixture (e.g., 75% argon and 25% CO_2) forms between the filler wire and the base metal.

Miller A brand of welding equipment. Miller welding machines are blue.

open-root joint A weld made in a space between two pieces of metal. The space is called the root opening.

oxyacetylene welding A welding process that uses oxygen and the fuel gas acetylene. Oxyacetylene welding is also called oxy-fuel welding or acetylene welding.

pad of beads A series of welds created to practice laying straight and consistent welds.

plasma cutting A melting process that uses a jet of ionized gas.

porosity A weld defect caused by gas trapped in the weld bead. The trapped gas generates pores the weaken the solidified weld.

PSI The abbreviation for pounds per square inch. In SMAW, the first two digits of an electrode number refer to tensile strength, measured in PSI. So, the 70 of a 7018 electrode means that the tensile strength of the weld bead will be 70,000 pounds per square inch.

root opening A space at the joint between two pieces of metal.

slag A byproduct of some welding processes, namely, SMAW and FCAW. In such processes, a flux coating shields the weld pool and prevents contamination. Slag forms after welding when the flux solidifies.

stick welding A welding process in which a welder creates an arc by striking a flux-covered electrode to the base metal. Stick welding is also called SMAW.

stinger The electrode holder in SMAW.

T-joint The welded point of two pieces of metal, one upright on top of the other, forming a 90-degree angle on either side and forming a T.

tack A small weld meant to hold pieces of metal together temporarily so that the welder can perform the final welds more easily and effectively.

TIG welding An abbreviation for tungsten-inert gas welding. TIG welding uses a tungsten electrode to create an electric arc and is also known as GTAW.

travel angle A relationship between the electrode and the work piece, measured perpendicular to the weld.

undercut A weld defect characterized by a groove in the base metal at the toe of the weld. Undercut has several causes, including improper technique.

voltage The force by which electrons pass through a conductor.

welpers A welding-specific type of pliers. Welders use them, for example, to hold hot metal, to pull and cut wire, and to clean spatter from the inside of a nozzle.

work angle A relationship between the electrode and the work piece, measured horizontally from the weld.

Notes

Introduction

1. In welding, a tack is a small weld meant to hold pieces of metal together temporarily so that the welder can perform the final welds more easily and effectively.

2. Gas-tungsten arc welding, or GTAW, is more commonly known as TIG (tungsten–inert gas) welding. GTAW uses a tungsten electrode to create an electric arc.

3. Shielded-metal arc welding, or SMAW, is more commonly known as stick welding. In SMAW, a welder creates an arc by striking a flux-covered electrode to the base metal.

4. I use a pseudonym for each person in this study. I use names that begin with T for the teachers. I use names that begin with S for the students. The only exception to this rule is for Ryan, which is the real name of my friend at DMACC. He told me I could use his name.

5. The first two digits of an electrode number refer to tensile strength. So, the 70 of a 7018 electrode means that the tensile strength of the weld bead will be 70,000 pounds per square inch, or PSI. The third number refers to the welding positions in which the electrode can be used. The 1 in 7018 means that the electrode can be used in all positions. The fourth number refers to the electrode's coating and to the current that can be used. The 8 in 7018 refers to low-hydrogen potassium and iron powder. It also indicates the electrode can be used with AC, DC positive, or DC negative current.

6. Welpers are a welding-specific type of pliers. Welders use them to hold hot metal, pull and cut wire, and clean spatter from the inside of a nozzle, among other things.

7. For a bend test, a student cuts two 1½-inch cross sections from their weld. One section is used to test the face side of the weld, and the other is used to test the root side. Each section is bent in a jig. The idea is to assess the weld's ductility and resistance to fracture.

8. An explanation of the transcription methods appears in chapter 3.

Chapter 1

1. Yet another study of gesture in technical communication is Sauer's (2003) analysis of miners' gestures as they recounted their experiences with risk in the workplace. In particular, Sauer revealed in chapters 7 and 8 how mine workers' gestures combined with their spoken narratives to describe the risks the workers had encountered and to generate new representations and viewpoints. Her participants, she wrote, "understand experience in and through gesture," as "spoken language does not easily accommodate the temporal and spatial complexity or the manner and motion that can be accommodated in gesture" (282). While Sauer's study delved deeply into the role of gesture in participants' narratives, the gestures it examined were those of mine workers *recollecting* their encounters with risk. It was not, in other words, a study of miners' use of gesture in the workplace itself.

2. The three major brands of welding machines are Miller (blue machines), Lincoln (red machines), and ESAB (yellow machines).

3. GMAW stands for gas-metal arc welding, also known as MIG (metal-inert gas) welding. GMAW passes metal filler wire from a spool through the welding gun, or torch. An electric arc, shielded by a gas mixture (e.g., 75% argon and 25% CO_2), forms between the filler wire and the base metal.

4. I discuss conventional and novel metaphors in more depth in chapter 3.

Chapter 2

1. MCC did not have an IRB.

2. I used this informed consent form for teachers as well.

3. Every welding student buys a grinder. Welders use grinders to grind out bad welds and to put a finish on welds. I also used my grinder at the end of each class period to remove stuck-on debris from my weld table.

Chapter 3

1. "Stinger" is welding jargon for the electrode holder in SMAW.

2. Work angle is measured horizontally from the weld; travel angle is measured perpendicular to the weld.

3. As he used describing strategies, Tom encountered one of many dangers involved with welding: a spark in the ear. He was ok.

4. Students make a pad of beads—row after row of welds—to practice laying straight and consistent welds. As they do, they also practice stopping and restarting a weld bead.

5. Structures on Earth's moon that look like dried-up riverbeds are not really old riverbeds. They are instead rilles—long, narrow channels generated by lava flows or sunken crust.

6. Slag is a byproduct of some welding processes, namely, SMAW and FCAW. It forms after welding when the flux solidifies.

7. I describe my (nerve-wracked) experience with this type of weld joint in chapter 5.

Chapter 4

1. FCAW stands for flux-core arc welding. FCAW differs from GMAW in that the gas shielding the arc comes from a flux inside the filler wire.

2. The other two of the traditional five senses—smell and taste—appear to play less of a role in welding students' development. In particular, it's fair to say that welding students did not use their sense of taste in the lab. However, welding certainly does produce a smell. In trying to describe the smell of space, astronaut Tony Antonelli reported that it smells like welding fumes, having "a distinct odor of ozone, a faint acrid smell" (Garber 2012). The fumes generated during different welding processes, such as the ozone formed during GMAW and GTAW, cause a range of health problems from irritating to life threatening. But in a workspace set up properly—with a fume extractor in each booth—the risks are mitigated. Other smells that I noticed while welding did not require olfactory expertise to identify. For example, I noticed the smell of burning rubber when I realized that I had stepped upon a small circle of metal, hot from being plasma cut. The metal had burned into the sole of my work boot. But I did not need to be an expert in welding to recognize the smell.

3. Teachers and students ask "ready?" when someone is observing them weld to ensure that the other person has lowered the face shield on their welding helmet.

4. More recently, research on the role of acoustics in welding has focused on the use of sound as an indicator of weld quality during automated welding processes (e.g., Lv et al. 2013; Zhu et al. 2019).

Chapter 5

1. Of course, the potential to earn a good wage is also an important extrinsic motivator for most welding students.

2. See Blum-Kulka, House, and Kaspern (1989) for a full description of downgraders, including the understaters "a little" and "a bit" (283–85).

3. Many women use men's welding gloves at first because they are readily available at welding suppliers. But even men's size small is too big for many

women. When I finally found gloves made for women, my welding life improved tremendously. The same is true for when I found "cheater" lenses to place into my welding helmet. All of a sudden, I could see the puddle clearly. (I also realized that I needed reading glasses.)

4. This counter to students' expectations of failure recalls another instance of Teresa's use of the giving sympathy strategy, an instance discussed in more detail in excerpt 5.4. In this interaction, Teresa made clear that learning how to weld a horizontal open-root joint in SMAW would take a few days:

TERESA: And just be ok that it- I mean it's going to take you two- at least two classes.
SOPHIA: Ok.
TERESA: You know. If you ask your fellow students they might tell you it took them even longer so don't- don't get-
SOPHIA: Ok.

As part of giving sympathy, Teresa explicitly pointed to other students' struggle with the task. But with this sympathy, Teresa implied these students' eventual success, just as the teacher in Vehkakoski's (2019) study explicitly pointed to other students' success as evidence that the student should feel optimistic.

5. Another term for "upgrader" is "intensifier" (e.g., Alcón-Soler, Safont Jordà, and Martínez-Flor 2005; Trosborg 1995).

6. Yes, I did in fact sell out my ancestors and Poles in general in my attempt to get a laugh.

7. Check out the subreddit Bad Welding to see how humor helps build an idioculture. This group identifies and pokes fun at bad welds.

8. In excerpt 3.6, see, for example, the discussion of Tonya's use of a novel metaphor that used a beer belly as the source domain.

9. Stein certainly was not the first one to run into this problem. Right before demonstrating a horizontal weld in SMAW, Teresa used humor when she pointed out the danger of failing to remove the filler-rod spacer:

TERESA: And again I like to take out the spacer before I tack the other side so the spacer doesn't get stuck in there.
SOPHIA: [Laughs]
TERESA: Then you have to pull it. Spend 15 minutes trying to get it out.
SOPHIA: [Laughs]

References

Abel, M. H. 1998. Interaction of Humor and Gender in Moderating Relationships between Stress and Outcomes. *The Journal of Psychology*, *132*(3), 267–76.

Abel, M. H. 2002. Humor, Stress, and Coping Strategies. *Humor*, *15*(4), 365–81.

Aghajanian, L. 2018. There's a Shortage of Welders. Will More Women Fill the Gap? *The Atlantic*. Retrieved from https://www.theatlantic.com/business/archive/2018/08/theres-a-shortage-of-welders-will-more-women-fill-the-gap/567434/

Alcón-Soler, E., Safont Jordà, P., and Martínez-Flor, A. 2005. Towards a Typology of Modifiers for the Speech Act of Requesting: A Socio-Pragmatic Approach. *RAEL: Revista Electrónica de Lingüística Aplicada*, *4*, 1–35.

Alexander, K., Schubert, A., and Meng, M. 2016. Does Detail Matter? The Effect of Visual Detail in Line Drawings on Task Execution. *Information Design Journal*, *22*(1), 49–61.

Andersen, P. A. 1999. *Nonverbal Communication: Forms and Functions*. Mountain View, CA: Mayfield.

Andersson, J., Öhman, M., and Garrison, J. 2018. Physical Education Teaching as a Caring Act—Techniques of Bodily Touch and the Paradox of Caring. *Sport, Education and Society*, *23*(6), 591–606.

Arata, Y., Inoue, K., Futamata, M., and Toh, T. 1979. Investigation on Welding Arc Sound (Report I): Effect of Welding Method and Welding Condition of Welding Arc Sound (Welding Physics, Processes & Instruments). *Transactions of the Joining and Welding Research Institute*, *8*(1), 25–31.

Arminen, I., Koskela, I., and Palukka, H. 2014. Multimodal Production of Second Pair Parts in Air Traffic Control Training. *Journal of Pragmatics*, *65*, 46–62.

Asplund, S., and Kilbrink, N. 2018. Learning How (and How Not) to Weld: Vocational Learning in Technical Vocational Education. *Scandinavian Journal of Educational Research*, *62*(1), 1–16.

Baake, K. 2003. *Metaphor and Knowledge: The Challenges of Writing Science*. Albany, NY: SUNY Press.

Bahl, E. K., Figueiredo, S., and Shivener, R. 2020. Comics and Graphic Storytelling in Technical Communication. *Technical Communication Quarterly, 29*(3), 219–21.

Bandura, A. 1977. Self-Efficacy: Toward a Unifying Theory of Behavioral Change. *Psychological Review, 84*(2), 191–215.

Bandura, A. 1986. *Social Foundations of Thought and Action: A Social Cognitive Theory*. Englewood Press, NJ: Prentice-Hall.

Bandura, A. 1997. *Self-Efficacy: The Exercise of Control*. New York: W. H. Freeman and Company.

Beckman, R. D. 2014. *The Effects of Audiation on the Melodic Error Detection Abilities of Fourth and Fifth Grade Band Students* (Doctoral dissertation). Retrieved from RUcore: Rutgers University Community Repository. https://doi.org/doi:10.7282/TeresaHD7X89

Becvar, A., Hollan, J., and Hutchins, E. 2008. Representational Gestures as Cognitive Artifacts for Developing Theories in a Scientific Laboratory. In M. S. Ackerman, C. A. Halverson, T. Erickson, and W. A. Kellogg (eds.), *Resources, Co-evolution and Artifacts: Theory in CSCW* (117–43). London: Springer-Verlag.

Berendt, E. A. 2008. Intersections and Diverging Paths. In E. A. Berendt (ed.), *Metaphors for Learning: Cross-Cultural Perspectives* (73–102). Netherlands: John Benjamins.

Berk, R. 1996. Student Ratings of Ten Strategies for Using Humor in College Teaching. *Journal on Excellence in College Teaching, 7*(3), 71–92.

Billett, S. 2000. Performance at Work: Identifying the Smart Workforce. In R. Gerber and C. Lankshear (eds.), *Training for a Smart Workforce* (123–50). London: Routledge.

Billett, S. 2001. Knowing in Practice: Re-Conceptualizing Vocational Expertise. *Learning and Instruction, 11*(6), 421–52.

Billett, S. 2002. Critiquing Workplace Learning Discourses: Participation and Continuity at Work. *Studies in the Education of Adults, 34*(1), 56–67.

Billett, S. 2015. Readiness and Learning in Health Care Education. *The Clinical Teacher, 12*(6), 367–72.

Billett, S. 2016. Learning Through Health Care Work: Premises, Contributions, and Practices. *Medical Education, 50*(1), 124–31.

Blum-Kulka, S., House, J., and Kasper, G. 1989. *Cross-Cultural Pragmatics: Requests and Apologies*. Norwood, NJ: Ablex.

Boerner, M., Joseph, S., and Murphy, D. 2017. The Association between Sense of Humor and Trauma-Related Mental Health Outcomes: Two Exploratory Studies. *Journal of Loss and Trauma, 22*(5), 440–52.

Bolkan, S., and Griffin, D. J. 2018. Catch and Hold: Instructional Interventions and Their Differential Impact on Student Interest, Attention, and Autonomous Motivation. *Communication Education, 67*(3), 269–86.

Booth-Butterfield, M., Booth-Butterfield, S., and Wanzer, M. 2007. Funny Students Cope Better: Patterns of Humor Enactment and Coping Effectiveness. *Communication Quarterly*, *55*(3), 299–315.

Booth-Butterfield, S., and Booth-Butterfield, M. 1991. Individual Differences in the Communication of Humorous Messages. *Southern Communication Journal*, *56*(3), 32–40.

Bourdieu, P. 1984. *Distinction: A Social Critique of the Judgement of Taste*. (R. Nice, trans.). Cambridge, MA: Harvard University Press.

Brand, G., and Brisson, R. 2012. Lateralisation in Wine Olfactory Threshold Detection: Comparison between Experts and Novices. *Laterality: Asymmetries of Body, Brain and Cognition*, *17*(5), 583–96.

Brophy, J. 1981. Teacher Praise: A Functional Analysis. *Review of Educational Research*, *51*(1), 5–32.

Brophy, J. 1983. Conceptualizing Student Motivation. *Educational Psychologist*, *18*(3), 200–15.

Brown, P., and Levinson, S. C. 1987. *Politeness: Some Universals in Language Usage*. UK: Cambridge University Press.

Brown, W., and Tomlin, J. 1996. Best and Worst University Teachers: The Opinion of Undergraduate Students. *College Student Journal*, *30*(1), 431–34.

Bryant, J., Comisky, P., and Zillman, D. 1997. Teachers' Humor in the College Classroom. *Communication Education*, *28*(2), 110–18.

Bsd0323. 2019. Response to u/sdawn13. [Online forum post]. Reddit. https://www.reddit.com/r/Welding/comments/ce7xy5/i_21f_am_considering_going_to_trade_school_for/

Byo, J. L., and Schlegel, A. L. 2016. Effects of Stimulus Octave and Timbre on the Tuning Accuracy of Advanced College Instrumentalists. *Journal of Research in Music Education*, *64*(3), 344–59.

Byo, J. L., and Sheldon, D. A. 2000. The Effect of Singing While Listening on Undergraduate Music Majors' Ability to Detect Pitch and Rhythm Errors. *Journal of Band Research*, *36*(1), 26–46.

Caldeborg, A., Maivorsdotter, N., and Öhman, M. 2019. Touching the Didactic Contract—a Student Perspective on Intergenerational Touch in PE. *Sport, Education and Society*, *24*(3), 256–68.

Cameron, L. 2003. *Metaphor in Educational Discourse*. London: Continuum.

Cameron, L., and Deignan, A. 2003. Using Large and Small Corpora to Investigate Tuning Devices Around Metaphor in Spoken Discourse. *Metaphor and Symbol*, *18*(3), 149–60.

Carter, S., and Pitcher, R. 2010. Extended Metaphors for Pedagogy: Using Sameness and Difference. *Teaching in Higher Education*, *15*(5), 579–89.

Cazden, C. B. 2001. *Classroom Discourse: The Language of Teaching and Learning*. Portsmouth, NH: Heinemann Educational Books.

Celaya, S. 2018. 5 Important Tips for Women in Welding. *BestWeldingGear*. Retrieved from https://www.bestweldinggear.com/5-important-tips-women-in-welding/

Cheung, C.-K., and Yue, D. Y. 2012. Sojourn Students' Humor Styles as Buffers to Achieve Resilience. *International Journal of Intercultural Relations, 36*, 353–64.

Chi, M. T. H., Siler, S. A. Jeong, H., Yamauchi, T., and Hausmann, R. G. 2001. Learning from Human Tutoring. *Cognitive Science 25*(4), 471–533.

cj737. 2018. Re: Body positioning and breathing. [Msg 3]. Message posted to https://forum.weldingtipsandtricks.com/viewtopic.php?f=2&t=13406

Clark, J. M., and Paivio, A. 1987. A Dual Coding Perspective on Encoding Processes. In M. A. McDaniel and M. Pressley (eds.), *Imagery and Related Mnemonic Processes: Theories, Individual Differences, and Applications* (5–33). New York: Springer.

Clark, J. M., and Paivio, A. 1991. Dual Coding Theory and Education. *Educational Psychology Review, 3*(3), 149–210.

Collins, A. 1991. Cognitive Apprenticeship and Instructional Technology. In L. Idol and B. F. Jones (eds.), *Educational Values and Cognitive Instruction: Implications for Reform* (121–38). Hillsdale, NJ: Lawrence Erlbaum.

Coulthard, M. 1985. *An Introduction to Discourse Analysis*. London: Routledge.

Coulthard, M. 2014. *An Introduction to Discourse Analysis: New edition*. London: Routledge.

Croijmans, I., and Majid, A. 2016. Not All Flavor Expertise is Equal: The Language of Wine and Coffee Experts. *PLoS One, 11*(6). https://doi.org/10.1371/journal.pone.0155845

Cromley, J. G., and Azevedo, R. 2005. What Do Reading Tutors Do? A Naturalistic Study of More and Less Experienced Tutors in Reading. *Discourse Processes, 40*(2), 83–113.

Croxton, R. A. 2014. The Role of Interactivity in Student Satisfaction and Persistence in Online Learning. *Journal of Online Learning and Teaching, 10*(2), 314–25.

Dan. 2004. Re: re: Sound of bacon frying. [Msg 3]. Message posted to https://weldtalk.hobartwelders.com/forum/weld-talk-topic-archive/welding-processes/7153-sound-of-bacon-frying

Deignan, A. 2011. Deliberateness is Not Unique to Metaphor: A Response to Gibbs. *Metaphor and the Social World, 1*(1), 57–60.

Demjén, Z. 2018. Complexity Theory and Conversational Humour: Tracing the Birth and Decline of a Running Joke in an Online Cancer Support Community. *Journal of Pragmatics, 133*, 93–104.

de Saint-Georges, I., and Filliettaz, L. 2008. Situated Trajectories of Learning in Vocational Training Interactions. *European Journal of Psychology of Education, 23*(2), 213–33.

Dix, S. 2016. Teaching Writing: A Multilayered Participatory Scaffolding Practice. *Literacy, 50*(1), 23–31.

Dodge, B. J., and Rossett, A. 1982. Heuristics for Humor in Instruction. *Performance and Instruction, 21*(4), 11–14.

Dörnyei, Z. 2001. New Themes and Approaches in Second Language Motivation Research. *Annual Review of Applied Linguistics, 21*, 43–59.

Eastwood. n.d. Watch and Listen to Your MIG Welder. It's Telling You Something! Retrieved from https://garage.eastwood.com/eastwood-chatter/listen-to-your-mig/

Ellis, C. 2004. *The Ethnographic I: A Methodological Novel about Autoethnography*. Walnut Creek, CA: Altamira.

Faber, P. (ed.). 2012. *A Cognitive Linguistics View of Terminology and Specialized Language*. Berlin: Walter de Gruyter.

Fan, M., Shi, S., and Truong, K. N. 2020. Practices and Challenges of Using Think-Aloud Protocols in Industry: An International Survey. *Journal of Usability Studies, 15*(2), 85–102.

Ferreira, M. M., and Bosworth, K. 2001. Defining Caring Teachers: Adolescents' Perspectives. *The Journal of Classroom Interaction, 36*(1), 24–30.

Filliettaz, L. 2010. Interactions and Miscommunication in the Swiss Vocational Education Context: Researching Vocational Learning from a Linguistic Perspective. *Journal of Applied Linguistics and Professional Practice, 7*(1), 27–50.

Filliettaz, L. 2011. Collective Guidance at Work: A Resource for Apprentices? *Journal of Vocational Education & Training, 63*(3), 485–504.

Filliettaz, L. 2013. Affording Learning Environments in Workplace Contexts: An Interactional and Multimodal Perspective. *International Journal of Lifelong Education, 32*(1), 107–22.

Fine, G. A. 1987. *With the Boys: Little League Baseball and Preadolescent Culture*. University of Chicago Press.

Fine, G. A., and De Soucey, M. 2005. Joking Cultures: Humor Themes as Social Regulation in Group Life. *Humor, 18*(1), 1–22.

Ford, T. E., Woodzicka, J. A., Triplett, S. R., Kochersberger, A. O., and Holden, C. J. 2014. Not All Groups are Equal: Differential Vulnerability of Social Groups to the Prejudice-Releasing Effects of Disparagement Humor. *Group Processes & Intergroup Relations, 17*(2), 178–99.

Fountain, T. K. 2014. *Rhetoric in the Flesh: Trained Vision, Technical Expertise, and the Gross Anatomy Lab*. London: Routledge.

Fuller, A., and Unwin, L. 2003. Learning as Apprentices in the Contemporary UK Workplace: Creating and Managing Expansive and Restrictive Participation. *Journal of Education and Work, 16*(4), 407–26.

Furnham, A. 1997. The Half Full or Half Empty Glass: The Views of the Economic Optimist vs Pessimist. *Human Relations, 50*(2), 197–209.

Furuyama, N. 2000. Gestural Interaction between the Instructor and the Learner in Origami Instruction. In D. McNeill (ed.), *Language and Culture* (99–117). UK: Cambridge University Press.

Garber, M. 2012. What Space Smells Like. *The Atlantic*. Retrieved from https://www.theatlantic.com/technology/archive/2012/07/what-space-smells-like/259903/

Gardner, R. C. 2007. Motivation and Second Language Acquisition. *Porta Linguarum*, *8*(3), 9–20.

Garner, R. L. 2006. Humor in Pedagogy: How Ha-Ha Can Lead to Aha! *College Teaching*, *54*(1), 177–80.

Garza, R. 2009. Latino and White High School Students' Perceptions of Caring Behaviors: Are We Culturally Responsive to Our Students? *Urban Education*, *44*(3), 297–321.

Gibbons, P. 2002. *Scaffolding Language, Scaffolding Learning: Teaching Second Language Learners in the Mainstream Classroom*. Portsmouth, NH: Heinemann.

Gilbert, A. N., Crouch, M., and Kemp, S. E. 1998. Olfactory and Visual Mental Imagery. *Journal Mental Imagery*, *22*, 137–46.

Giles, T. D. 2008. *Motives for Metaphor in Scientific and Technical Communication*. Amityville, NY: Baywood.

Glynn, M. A. 1994. Effects of Work Task Cues and Play Task Cues on Information Processing, Judgment, and Motivation. *Journal of Applied Psychology*, *79*(1), 34–45.

Goetz, T., and Hall, N. C. 2014. Academic Boredom. In R. Pekrun, and L. Linnenbrink-Garcia (eds.), *International Handbook of Emotions in Education* (311–30). London: Routledge.

Goh, C. C., and Hu, G. 2014. Exploring the Relationship between Metacognitive Awareness and Listening Performance with Questionnaire Data. *Language Awareness*, *23*(3), 255–74.

Gonzo, C. L. 1971. An Analysis of Factors Related to Choral Teachers' Ability to Detect Pitch Errors While Reading the Score. *Journal of Research in Music Education*, *19*(3), 259–71.

Goodwin, C. 1994. Professional Vision. *American Anthropologist*, *96*(3), 606–33.

Goschler, J. 2019. Metaphors in Educational Texts: A Case Study on History and Chemistry Teaching Material. *Yearbook of the German Cognitive Linguistics Association*, *7*(1), 79–92.

Grady, J. 2017. Using Metaphor to Influence Public Perceptions and Policy: How Metaphors can Save the World. In E. Semino and Z, Demjén (eds.), *The Routledge Handbook of Metaphor and Language* (443–54). London: Taylor & Francis.

Graesser, A. C., Person, N. K., and Magliano, J. P. 1995. Collaborative Dialogue Patterns in Naturalistic One-to-One Tutoring. *Applied Cognitive Psychology* *9*(6), 495–522.

Gullberg, M. 1998. *Gesture as a Communication Strategy in Second Language Discourse: A Study of Learners of French and Swedish*. Sweden: Lund University Press.

Gullberg, M. 2006. Some Reasons for Studying Gesture and Second Language Acquisition (Hommage à Adam Kendon). *International Review of Applied Linguistics in Language Teaching*, *44*, 103–24.

Gullberg, M. 2008. A Helping Hand? Gestures, L2 Learners, and Grammar. In S. G. McCafferty and G. Stam (eds.), *Gesture, Second Language Acquisition and Classroom Research* (185–210). London: Routledge.

Gunney. 2013. Response to "Aluminum MIG help needed!" [Msg 2]. Message posted to https://forum.millerwelds.com/forum/welding-discussions/32501-aluminum-mig-help-needed

Haas, C., and Witte, S. P. 2001. Writing as an Embodied Practice: The Case of Engineering Standards. *Journal of Business and Technical Communication*, *15*(4), 413–57.

Hall, R. V., Lund, D., and Jackson, D. 1968. Effects of Teacher Attention on Study Behavior 1. *Journal of Applied Behavior Analysis*, *1*(1), 1–12.

Hermkes, R., Mach, H., and Minnameier, G. 2018. Interaction-Based Coding of Scaffolding Processes. *Learning and Instruction*, *54*, 147–55.

Hidi, S., and Boscolo, P. 2006. Motivation and Writing. In C. A. MacArthur, S. Graham, and J. Fitzgerald (eds.), *Handbook of Writing Research* (144–57). New York: Guilford.

Hindmarsh, J., and Pilnick, A. 2007. Knowing Bodies at Work: Embodiment and Ephemeral Teamwork in Anaesthesia. *Organization Studies*, *28*(9), 1395–416.

Hindmarsh, J., Reynolds, P., and Dunne, S. 2011. Exhibiting Understanding: The Body in Apprenticeship. *Journal of Pragmatics*, *43*(2), 489–503.

Horn, H. L., and Maxwell, F. R. 1983. The Impact of Task Difficulty Expectations on Intrinsic Motivation. *Motivation and Emotion*, *7*(1), 19–24.

Ingold, T. 2001. From the Transmission of Representations to the Education of Attention. In H. Whitehouse (ed.), *The Debated Mind* (113–53). Oxford, UK: Berg.

Jeffrey Grady. 2008. Response to "Wondering if I should choose a career in welding." [Msg 6]. Message posted to https://app.aws.org/forum/topic_show.pl?tid=15521

Jin, L., and Cortazzi, M. 2011. More Than a Journey: "Learning" in the Metaphors of Chinese Students and Teachers. In L. Jin and M. Cortazzi (eds.), *Researching Chinese Learners* (67–92). London: Palgrave Macmillan.

Jones, S. H., Adams, T. E., and Ellis, C. (eds.) 2016. *Handbook of Autoethnography*. London: Routledge.

Jussim, L., and Harber, K. D. 2005. Teacher Expectations and Self-Fulfilling Prophecies: Knowns and Unknowns, Resolved and Unresolved Controversies. *Personality and Social Psychology Review*, *9*(2), 131–55.

Karami, A., Kahrazei, F., and Arab, A. 2018. The Role of Humor in Hope and Posttraumatic Growth among Patients with Leukemia. *Journal of Fundamentals of Mental Health*, *20*(3), 176–84.

Kendon, A. 2000. Language and Gesture: Unity or Duality? In D. McNeill (ed.), *Language and Gesture* (47–63). UK: Cambridge University Press.

Kendon, A. 2004. *Gesture: Visible Action as Utterance*. UK: Cambridge University Press.

Kim, S., and Cho, S. 2017. How a Tutor Uses Gesture for Scaffolding: A Case Study on L2 Tutee's Writing. *Discourse Processes*, *54*(2), 105–23.

Kita, S. 2000. How Representational Gestures Help Speaking. In D. McNeill (ed.), *Language and Gesture* (162–85). UK: Cambridge University Press.

Kita, S. (ed.). 2003. *Pointing: Where Language, Culture, and Cognition Meet*. Mahwah, NJ: Lawrence Erlbaum Associates.

Korobkin, D. 1989. Humor in the Classroom: Considerations and Strategies. *College Teaching*, *36*(4), 154–58.

Koschmann, T., LeBaron, C., Goodwin, C., and Feltovich, P. 2011. "Can You See the Cystic Artery Yet?": A Simple Matter of Trust. *Journal of Pragmatics*, *43*, 521–41.

Kostelnick, C. 2016. The Re-Emergence of Emotional Appeals in Interactive Data Visualization. *Technical Communication*, *63*(2), 116–35.

Kotthoff, H. 2009. Joint Construction of Humorous Fictions in Conversation. An Unnamed Narrative Activity in a Playful Keying. *Journal of Literary Theory*, *3*(2), 195–217.

Kövecses, Z. 2010. *Metaphor: A Practical Introduction*. UK: Oxford University Press.

Kuiper, N. A., and Martin, R. A. 1998. Is Sense of Humor a Positive Personality Characteristic? In W. Ruch (ed.), *The Sense of Humor: Explorations of a Personality Characteristic* (159–78). Berlin: Mouton de Gruyter.

Lakoff, G. 2006. Conceptual Metaphor. In D. Geeraerts (ed.), *Cognitive Linguistics: Basic Readings* (185–239). Berlin: Walter de Gruyter.

Lakoff, G., and Johnson, M. 1980. *Metaphors We Live By*. Chicago University Press.

Lave, J., and Wenger, E. 1991. *Situated Learning: Legitimate Peripheral Participation*. UK: Cambridge University Press.

Lazaraton, A. 2004. Gesture and Speech in the Vocabulary Explanations of One ESL Teacher: A Microanalytic Inquiry. *Language Learning*, *54*(1), 79–117.

Lepper, M. R., Woolverton, M., Mumme, D. L., and Gurtner, J.-L. 1993. Motivational Techniques of Expert Human Tutors: Lessons for the Design of Computer-Based Tutors. In S. P. Lajoie (ed.), *Computers as Cognitive Tools: Technology in Education* (75–105). Hillsdale, NJ: Lawrence Erlbaum.

Li, L. 2020. Visualizing Chinese Immigrants in the US Statistical Atlases: A Case Study in Charting and Mapping the Other(s). *Technical Communication Quarterly*, *29*(1), 1–17.

Littlemore, J., and Low, G. D. 2006a. *Figurative Thinking and Foreign Language Learning*. London: Springer.

Littlemore, J., and Low, G. 2006b. Metaphoric Competence, Second Language Learning, and Communicative Language Ability. *Applied Linguistics*, *27*(2), 268–94.

Livermore, A., and Laing, D. G. 1996. Influence of Training and Experience on the Perception of Multicomponent Odor Mixtures. *Journal of Experimental Psychology: Human Perception and Performance*, *22*(2), 267–77.

Luthans, B. C., Luthans, K. W., and Jensen, S. M. 2012. The Impact of Business School Students' Psychological Capital on Academic Performance. *Journal of Education for Business*, *87*(5), 253–59.

Lv, N., Xu, Y., Zhang, Z., Wang, J., Chen, B., and Chen, S. 2013. Audio Sensing and Modeling of Arc Dynamic Characteristic During Pulsed AI Alloy GTAW Process. *Sensor Review*, *33*(2), 141–56.

Machlev, M., and Karlin, N. J. 2017. The Relationship between Instructor Use of Different Types of Humor and Student Interest in Course Material. *College Teaching*, *65*(4), 192–200.

Mackiewicz, J. 2006. The Functions of Formulaic and Nonformulaic Compliments in Interactions about Technical Writing. *IEEE Transactions on Professional Communication*, *49*(1), 12–27.

Mackiewicz, J., and Thompson, I. 2018. *Talk About Writing: The Tutoring Strategies of Experienced Writing Center Tutors* (2nd ed.). London: Routledge.

Madhyastha, S., Latha, K. S., and Kamath, A. 2014. Stress, Coping and Gender Differences in Third Year Medical Students. *Journal of Health Management*, *16*(2), 315–26.

Mainland, J. D., Bremner, E. A., Young, N., Johnson, B. N., Khan, R. M., Bensafi, M., and Sobel, N. 2002. Olfactory Plasticity: One Nostril Knows What the Other Learns. *Nature*, 419. https://doi: 10.1038/419802a

Many, J. E. 2002. An Exhibition and Analysis of Verbal Tapestries: Understanding how Scaffolding is Woven into the Fabric of Instructional Conversations. *Reading Research Quarterly*, *37*(4), 376–407.

Margolis, H. 2005. Increasing Struggling Learners' Self-Efficacy: What Tutors Can Do and Say. *Mentoring & Tutoring: Partnership in Learning*, *13*(2), 221–38.

Mazer, J. P. 2012. Development and Validation of the Student Interest and Engagement Scales. *Communication Methods and Measures*, *6*(2), 99–125.

McCafferty, S. G. 2002. Gesture and Creating Zones of Proximal Development for Second Language Learning. *The Modern Language Journal*, *862*(2), 192–203.

McCafferty, S. G. 2004. Space for Cognition: Gesture and Second Language Learning. *International Journal of Applied Linguistics*, *14*(1), 148–65.

McCafferty, S. G. 2006. Gesture and the Materialization of Second Language Prosody. *International Review of Applied Linguistics in Language Teaching*, *44*(2), 197–209.

McLain, M. 2018. Emerging Perspectives on the Demonstration as a Signature Pedagogy in Design and Technology Education. *International Journal of Technology and Design Education*, *28*(4), 985–1000.

McLaughlin, T. F. 1982. The Effects of Teacher Praise on Accuracy of Math Performance for an Entire Special Education Classroom. *Behavioral Engineering*, *7*(3), 81–86.

McNeill, D. 1992. *Hand and Mind: What Gestures Reveal about Thought*. University of Chicago Press.

McNeill, D. 2005. *Gesture and Thought*. University of Chicago Press.

McNeill, D. 2016. *Why We Gesture: The Surprising Role of Hand Movements in Communication*. UK: Cambridge University Press.

Medoro, C., Cianciabella, M., Camilli, F., Magli, M., Gatti, E., and Predieri, S. 2016. Sensory Profile of Italian Craft Beers, Beer Taster Expert Versus Sensory Methods: A Comparative Study. *Food and Nutrition Sciences*, *7*(6), doi: 10.4236/fns.2016.76047

Melander, H., and Sahlström, F. 2009. In Tow of the Blue Whale: Learning as Interactional Changes in Topical Orientation. *Journal of Pragmatics*, *41*(8), 1519–37.

Mentis, H. M., Chellali, A., and Schwaitzberg, S. 2014. Learning to See the Body: Supporting Instructional Practices in Laparoscopic Surgical Procedures. In *Proceedings of the ACM CHI Conference on Human Factors in Computing Systems* (2113–22). Retrieved from https://hal.archives-ouvertes.fr/hal-00957806/document

Mercer, N. 1995. *The Guided Construction of Knowledge*. Bristol, UK: Multilingual Matters.

Merrill, D. C., Reiser, B. J., Merrill, S. K., and Landes, S. 1995. Tutoring: Guided Learning by Doing. *Cognition and Instruction 13*(3), 315–72.

Miles, E. M. 1972. Beat Elimination as a Means of Teaching Intonation to Beginning Wind Instrumentalists. *Journal of Research in Music Education*, *20*(4), 496–500.

Mondada, L. 2012. Video Analysis and the Temporality of Inscriptions within Social Interaction: The Case of Architects at Work. *Qualitative Research*, *12*(3), 304–33.

Moskovsky, C., Alrabai, F., Paolini, S., and Ratcheva, S. 2013. The Effects of Teachers' Motivational Strategies on Learners' Motivation: A Controlled Investigation of Second Language Acquisition. *Language Learning*, *63*(1), 34–62.

Murray, H., and Lang, M. 1997. Does Classroom Participation Improve Student Learning? *Teaching and Learning in Higher Education*, *20*(1), 7–9.

Myhill, D., and Warren, P. 2005. Scaffolds or Straitjackets? Critical Moments in Classroom Discourse. *Educational Review*, *57*(1), 55–69.

Nevile, M. 2007. Action in Time: Ensuring Timeliness for Collaborative Work in the Airline Cockpit. *Language in Society*, *36*, 233–57.

Nielson, K. 2008. Scaffold Instruction at the Workplace From a Situated Perspective. *Studies in Continuing Education*, *30*(3), 247–61.

Noble, C., and Billett, S. 2017. Learning to Prescribe Through Co-Working: Junior Doctors, Pharmacists and Consultants. *Medical Education*, *51*(4), 442–51.

Öhman, M. 2017. Losing Touch–Teachers' Self-Regulation in Physical Education. *European Physical Education Review, 23*(3), 297–310.

O'Leary, K., and O'Leary, S. 1977. *Classroom Management: The Successful Use of Behavior Modification*. Oxford, UK: Pergamon Press.

Paivio, A. 1991. Dual Coding Theory: Retrospect and Current Status. *Canadian Journal of Psychology/Revue Canadienne de Psychologie, 45*(3), 255–87.

Paladin. 2011. Overhead welding. [Msg 5]. Message posted to https://app.aws.org/forum/topic_show.pl?tid=27874

Palincsar, A. S. 1986. The Role of Dialogue in Providing Scaffolded Instruction. *Educational Psychologist, 21*(1–2), 73–98.

Palmer, D. K. 2009. Code-Switching and Symbolic Power in a Second-Grade Two-Way Classroom: A Teacher's Motivation System Gone Awry. *Bilingual Research Journal, 32*(1), 42–59.

Papi, M., and Abdollahzadeh, E. 2012. Teacher Motivational Practice, Student Motivation, and Possible L2 Selves: An Examination in the Iranian EFL Context. *Language Learning, 62*(2), 571–94.

Park, I. 2015. Or-Prefaced Third Turn Self-Repairs in Student Questions. *Linguistics and Education, 31*, 101–14.

Park, I. 2017. Questioning as Advice Resistance: Writing Tutorial Interactions with L2 Writers. *Classroom Discourse, 8*(3), 253–70.

Pekrun, R., Goetz, T., Daniels, L. M., and Stupinsky, R. H. 2010. Boredom in Achievement Settings: Exploring Control-Value Antecedents and Performance Outcomes of a Neglected Emotion. *Journal of Educational Psychology, 102*(3), 531–49.

Person, N. K., Graesser, A. C., Magliano, J. P., and Kreuz, R. J. 1994. Inferring What the Student Knows in One-to-One Tutoring: The Role of Student Questions and Answers. *Learning and Individual Differences, 6*(2), 205–29.

Phan, H. P. 2016. Longitudinal Examination of Optimism, Personal Self-Efficacy and Student Well-Being: A Path Analysis. *Social Psychology of Education, 19*, 403–26.

Piemonte, N. M. 2015. Last Laughs: Gallows Humor and Medical Education. *Journal of Medical Humanities, 36*(4), 375–90.

Pinnell, G. S., Lyons, C. A., DeFord, D. E., Bryk, A. S., and Seltzer, M. 1994. Comparing Instructional Models for the Literacy Education of High-Risk First Graders. *Reading Research Quarterly 29*(1), 9-39.

Pollio, H. R., and Lee Humphreys, W. 1996. What Award-Winning Lecturers Say about Their Teaching: It's All about Connection. *College Teaching, 44*(3), 101–106.

Pratton, J., and Hales, L. W. 1986. The Effects of Active Participation on Student Learning. *The Journal of Educational Research, 79*(4), 210–15.

Puntambekar, S., and Hübscher, R. 2005. Tools for Scaffolding Students in a Complex Learning Environment: What Have We Gained and What Have We Missed? *Educational Psychologist, 40*(1), 1–12.

Qian, X., Meng, H., and Soong, F. K. 2012. The Use of DBN-HMMs for Mispronunciation Detection and Diagnosis in L2 English to Support Computer-Aided Pronunciation Training. In *Thirteenth Annual Conference of the International Speech Communication Association*. Retrieved from http://www1.se.cuhk.edu.hk/~hccl/publications/pub/xiaojun_interspeech2012.pdf

Rahman, A., Palaneeswaran, E., Kulkarni, A., and Zou, P. 2015. Musculoskeletal Health and Safety of Aged Workers in Manual Handling Works. In *2015 International Conference on Industrial Engineering and Operations Management* (IEOM) (1–4). IEEE.

Rawlins, J. D., and Wilson, G. D. 2014. Agency and Interactive Data Displays: Internet Graphics as Co-Created Rhetorical Spaces. *Technical Communication Quarterly*, *23*(4), 303–22.

Reddy, M. 1979. The Conduit Metaphor. *Metaphor and Thought*, *2*, 285–324.

Rodgers, J. V., D'Agostino, J. V., Harmey, S. J., Kelly, R. H., and Brownfield, K. 2016. Examining the Nature of Scaffolding in an Early Literacy Intervention. *Reading Research Quarterly*, *51*(3), 345–60.

Ross, D. G. 2017. The Role of Ethics, Culture, and Artistry in Scientific Illustration. *Technical Communication Quarterly*, *26*(2), 145–72.

Roth, W. M. 2002. From Action to Discourse: The Bridging Function of Gestures. *Cognitive Systems Research*, *3*(3), 535–54.

Rowe, A., and Regehr, C. 2010. Whatever Gets you Through Today: An Examination of Cynical Humor among Emergency Service Professionals. *Journal of Loss and Trauma*, *15*(5), 448–64.

Royet, J. P., Plailly, J., Saive, A. L., Veyrac, A., and Delon-Martin, C. 2013. The Impact of Expertise in Olfaction. *Frontiers in Psychology*, *4*, article 928.

Rude, C. D. 2009. Mapping the Research Questions in Technical Communication. *Journal of Business and Technical Communication*, *23*(2), 174–215.

Ruoranen, M., Antikainen, T., and Eteläpelto, A. 2017. Surgical Learning and Guidance on Operative Risks and Potential Errors. *Journal of Workplace Learning*, *29*(5), 326–42.

Sakai, S., Korenaga, R., Mizukawa, Y., and Igarashi, M. 2014. Envisioning the Plan in Interaction: Configuring Pipes During a Plumbers' Meeting. In M. Nevile, P. Haddington, T. Heinemann, and M. Rauniomaa (eds.), *Interacting with Objects: Language Materiality and Social Activity* (339–56). Netherlands: John Benjamins.

Sauer, B. 2003. *The Rhetoric of Risk: Technical Documentation in Hazardous Environments*. Hillsdale, NJ: Lawrence Erlbaum.

Scherber, R. V. 2014. *Pedagogical Practices Related to the Ability to Discern and Correct Intonation Errors: An Evaluation of Current Practices, Expectations, and a Model for Instruction* (doctoral dissertation). Retrieved from https://diginole.lib.fsu.edu/islandora/object/fsu%3A254498

Scherr, R. E. 2008. Gesture Analysis for Physics Education Researchers. *Physical Review Special Topics: Physics Education Research*, *4*, 1–9.

Schiffrin, D., Tannen, D., and Hamilton, H. E. (eds.) 2001. *The Handbook of Discourse Analysis*. Oxford, UK: Blackwell.

Sharpe, T. 2006. 'Unpacking' Scaffolding: Identifying Discourse and Multimodal Strategies that Support Learning. *Language and Education*, *20*(3), 211–31.

Sheldon, D. A. 2004. Effects of Multiple Listenings on Error-Detection Acuity in Multivoice, Multitimbral Musical Examples. *Journal of Research in Music Education*, *52*(2), 102–15.

Siegel, J. 2013. Exploring L2 Listening Instruction: Examinations of Practice. *ELT Journal*, *68*(1), 22–30.

Silvey, B. A., Nápoles, J., and Springer, D. G. 2019. Effects of Pre-Tuning Vocalization Behaviors on the Tuning Accuracy of College Instrumentalists. *Journal of Research in Music Education*, *66*(4), 392–407.

Singh, B., and Singhal, P. 2016. Work Related Musculoskeletal Disorders (WMSDS) Risk Assessment for Different Welding Positions and Processes. In L. P. Singh, S. Singh, and A. Bhardwaj (eds.), *14th International Conference on Humanizing Work and Work Environment HWWE* (264–67). Indian Society of Ergonomics and International Ergonomics Association.

Skarbø Solem, M. 2016. Displaying Knowledge through Interrogatives in Student-Initiated Sequences. *Classroom Discourse* *7*(1), 18–35.

Smit, J., van Eerde, H. A. A., and Bakker, A. 2013. A Conceptualisation of Whole-Class Scaffolding. *British Educational Research Journal*, *39*(5), 817–34.

Society for Technical Communication. 2020. Defining Technical Communication. Retrieved from https://www.stc.org/about-stc/defining-technical-communication/

Stam, G. 2006. Thinking for Speaking about Motion: L1 and L2 Speech and Gesture. *International Review of Applied Linguistics in Language Teaching*, *44*, 145–71.

Strik, H., Truong, K., De Wet, F., and Cucchiarini, C. 2009. Comparing Different Approaches for Automatic Pronunciation Error Detection. *Speech Communication*, *51*(10), 845–52.

Sutkin, G., Littleton, E. B., and Kanter, S. L. 2015a. How Surgical Mentors Teach: A Classification of In Vivo Teaching Behaviors Part 1: Verbal Teaching Guidance. *Journal of Surgical Education*, *72*(2), 243–50.

Sutkin, G., Littleton, E. B., and Kanter, S. L. 2015b. How Surgical Mentors Teach: A Classification of In Vivo Teaching Behaviors Part 2: Physical Teaching Guidance. *Journal of Surgical Education*, *72*(2), 251–57.

Swanton Welding Company. 2016. The Difference Between Defects and Discontinuities. [Blog post]. Retrieved from https://blog.swantonweld.com/welding-inspection-defects-discontinuities

Swift, V., and Peterson, J. B. 2018. Improving the Effectiveness of Performance Feedback by Considering Personality Traits and Task Demands. *PLOS ONE*, *13*(5). Retrieved from https://doi.org/10.1371/journal.pone.0197810

Syraphina. 2019. Response to "Do women actually have a hard time with men at their jobs?" [Msg 7]. Message posted to https://www.reddit.com/r/Welding/comments/e60eur/do_women_actually_have_a_hard_time_with_the_men/

Talbert, M. D. 2012. *Adult Amateur Musicians and Melodic Error Detection* (doctoral dissertation). Retrieved from https://scholarcommons.sc.edu/etd/1632

Tarn, J., and Huissoon, J. 2005. Developing Psycho-Acoustic Experiments in Gas Metal Arc Welding. In *IEEE International Conference Mechatronics and Automation* (1112–117). IEEE. Retrieved from https://doi.org/10.1109/ICMA.2005.1626707

Tetzner, J., and Becker, M. 2017. Think Positive? Examining the Impact of Optimism on Academic Achievement in Early Adolescents. *Journal of Personality*, *85*(3), 1–13.

The Welding Master. 2017. What is Welding Defects? Types, Causes, and Remedies? Retrieved from https://www.theweldingmaster.com/welding-defects/

Thompson, I. 2009. Scaffolding in the Writing Center: A Microanalysis of an Experienced Tutor's Verbal and Nonverbal Tutoring Strategies. *Written Communication, 26*(4), 417–53.

Thonus, T. 2010. Metaphorical Language in Writing Center Consultations. International Writing Centers Association, Baltimore, MD, 5 November.

Thonus, T., and Hewitt, B. L. 2016. Follow this PATH: Conceptual Metaphors in Writing Center Online Consultations. *Metaphor and the Social World*, *6*(1), 52–78.

Thornton, L. C. 2008. The Effect of Grade, Experience, and Listening Condition on the Melodic Error Detection of Fifth-and Sixth-Grade Woodwind Students. *Update: Applications of Research in Music Education*, *26*(2), 4–10.

Thuma, H., and Miranda, K. 2020. Hands On/Hands Off: Pedagogical Touch in the# MeToo Era. *Voice and Speech Review*, *14*(2), 213–26.

TimGary. 2002. Response to "Overhead SMAW." [Msg 2]. Message posted to https://app.aws.org/forum/topic_show.pl?tid=2576

Trosborg, A. 1995. *Interlanguage Pragmatics: Requests, Complaints, and Apologies*. Berlin: Mouton de Gruyter.

Turner, J. C., and Meyer, D. K. 2004. A Classroom Perspective on the Principle of Moderate Challenge in Mathematics. *The Journal of Educational Research*, *97*(6), 311–18.

Urhahne, D. 2015. Teacher Behavior as a Mediator of the Relationship between Teacher Judgment and Students' Motivation and Emotion. *Teaching and Teacher Education*, *45*, 73–82.

US Bureau of Labor Statistics. 2020a. Employed Persons by Detailed Occupation, Sex, Race, and Hispanic or Latino Ethnicity. Retrieved from https://www.bls.gov/cps/cpsaat11.htm

US Bureau of Labor Statistics. 2020b. Welders, Cutters, Solderers, and Brazers. Retrieved from https://www.bls.gov/ooh/production/welders-cutters-solderers-and-brazers.htm

van de Pol, J., and Elbers, E. 2013. Scaffolding Student Learning: A Micro-Analysis of Teacher–Student Interaction. *Learning, Culture and Social Interaction*, *2*(1), 32–41.

van de Pol, J., Volman, M., and Beishuizen, J. 2010. Scaffolding in Teacher-Student Interaction: A Decade of Research. *Educational Psychology Review*, *22*(3), 271–96.

van de Pol, J., Volman, M., and Beishuizen, J. 2012. Promoting Teacher Scaffolding in Small-Group Work: A Contingency Perspective. *Teaching and Teacher Education*, *28*(2), 193–205.

van Nispen, K., van de Sandt-Koenderman, W. M., and Krahmer, E. 2017. Production and Comprehension of Pantomimes Used to Depict Objects. *Frontiers in Psychology*, *8*, doi.org/10.3389/fpsyg.2017.01095

Van Praag, L., Stevens, P. A., and Van Houtte, M. 2017. How Humor Makes or Breaks Student-Teacher Relationships: A Classroom Ethnography in Belgium. *Teaching and Teacher Education*, *66*, 393–401.

Vandergrift, L., and Goh, C. C. 2012. *Teaching and Learning Second Language Listening: Metacognition in Action*. London: Routledge.

Vandergrift, L., and Tafaghodtari, M. H. 2010. Teaching L2 Learners How to Listen Does Make a Difference: An Empirical Study. *Language Learning*, *60*(2), 470–97.

Vehkakoski, T. M. 2019. "Can Do!": Teacher Promotion of Optimism in Response to Student Failure Expectation Expressions in Classroom Discourse. *Scandinavian Journal of Educational Research*, *63*, https://doi.org/10.1080/00313831.2019.1570547

Vibulphol, J. 2016. Students' Motivation and Learning and Teachers' Motivational Strategies in English Classrooms in Thailand. *English Language Teaching*, *9*(4), 64–75.

Wang, Q., and Spence, C. 2018. Assessing the Influence of Music on Wine Perception among Wine Professionals. *Food Science & Nutrition*, *6*(2), 295–301.

Wanzer, M. B., Frymier, A. B., Wojtaszczyk, A. M., and Smith, T. 2006. Appropriate and Inappropriate Uses of Humour by Teachers. *Communication Education*, *55*(2), 178–96.

Willox, A. C., Harper, S. L., Bridger, D., Morton, S., Orbach, A., and Sarapura, S. 2010. Co-Creating Metaphor in the Classroom for Deeper Learning: Graduate Student Reflections. *International Journal of Teaching and Learning in Higher Education*, *22*(1), 71–9.

Wilson, G. D., Rawlins, J. D., and Crane, K. 2018. Agency in Action: Exploring User Responses and Rhetorical Choices in Interactive Data Displays. *Journal of Technical Writing and Communication*, *48*(4), 471–99.

Witte, S. P. 1987. Pre-Text and Composing. *College Composition and Communication*, *38*(4), 397–425.

Wolfe, J. 2005. Gesture and Collaborative Planning: A Case Study of a Student Writing Group. *Written Communication*, *22*(3), 298–332.

Wood, D., Bruner, J. S., and Ross, G. 1976. The Role of Tutoring in Problem Solving. *Journal of Child Psychiatry and Psychology*, *17*(2), 89–100.

Wooffitt, R. 2005. *Conversation Analysis and Discourse Analysis: A Comparative and Critical Introduction*. Thousand Oaks, CA: SAGE.

Yu, H. 2016. *The Other Kind of Funnies: Comics in Technical Communication*. London: Routledge.

Zarzo, M., and Stanton, D. T. 2009. Understanding the Underlying Dimensions in Perfumers' Odor Perception Space as a Basis for Developing Meaningful Odor Maps. *Attention, Perception, & Psychophysics, 71*(2), 225–47.

Zhang, Y. 2016. Illustrating Beauty and Utility: Visual Rhetoric in Two Medical Texts Written in China's Northern Song Dynasty, 960–1127. *Journal of Technical Writing and Communication, 46*(2), 172–205.

Zhu, T., Shi, Y., Cui, S., and Cui, Y. 2019. Recognition of Weld Penetration During K-TIG Welding Based on Acoustic and Visual Sensing. *Sensing and Imaging, 20*(3). https://doi.org/10.1007/s11220-018-0224-9

Index